NEVER GET LOST

A GREEN BERET'S GUIDE TO LAND NAVIGATION

DAVID WALTON

Copyright © 2024 by David Walton

All rights reserved.

No part of this book may be reproduced in any form or by any electronic or mechanical means, including information storage and retrieval systems, without written permission from the author, except for the use of brief quotations in a book review.

Brad B gets credit for the cover and so much more...

❦ Created with Vellum

For The Brotherhood

*Good Men Will Not Just Sit and
Watch Other Men Suffer.*

*And to the boys of SFODA 715
John, Stew, Pat, Brad, Quincy, Cleaner, Cesar, Kevin,
Skip, Jerome, & Lee*

Thanks for letting me hold the map.

DOL

DISCLAIMER

Land navigation is hard. SFAS is hard. Life is hard. I make no guarantees about the results of the information applied in this book. I share educational and informational resources that are intended to help you succeed in SFAS and help you to land navigate. You need to know that your ultimate success or failure will be the result of your own efforts, your particular situation, and innumerable other circumstances beyond my knowledge and control. Your adequate preparation will require immense physical effort. You should consult with a physician to determine your suitability prior to any physical training.

This book is not endorsed by the Department of Defense, the United States Army Special Operations Command, or the United States Army John F Kennedy Special Warfare Center and School...but it should be.

PROLOGUE

Land navigation failures account for about half of Special Forces Assessment and Selection (SFAS) drops. There are a multitude of contributing factors, but basic knowledge and training deficiency are very high on the list and are the foundation upon which the rest of the factors rest. My company, TFVooDoo LLC, exists for the sole purpose of preparing the next generation of aspiring Green Berets for the rigors of SFAS, including land navigation. So while this book certainly can be valuable for anyone learning land navigation, it is specifically geared for SFAS prep. And because the land navigation at SFAS can be fairly limited, we will also cover lots of stuff that you might not necessarily use at SFAS, but we need to cover it so the stuff that you do use makes more sense.

Tell, Show, Do. This is a time-honored and proven teaching methodology; Tell the student how something works, Show them how it works, and then let them make the thing work. Tell, Show, Do. This is especially true with a tactile skill like land navigation. I believe that you have to get all the way to the "do" stage before it really starts to make sense. And even more directly, land navigation is one of the few skills that you must actually fail in order to fully learn it. So, we have to recognize that this book is the first step in the skills trilogy -- tell. We are going to *tell* you how this stuff works in this book. We have, in addition to the plentiful graphics in this book, a series of video tutorials that are designed to specifically *show* you the most critical skills. We'll show you how to shoot an azimuth and plot a point and the step by step written instructions from this book will immediately come together in a more meaningful way. And finally, TFVooDoo hosts in-person land navigation Muster events every month in the Sandhills of North Carolina, the ancestral birthplace of the Green Berets. There you can actually put into practice – *do* –the skills that we are telling you about here and showing you in the graphics and videos. Tell, show, do.

Don't expect to become a master navigator like Marco Polo or a spatial modeler like Nicolaus Copernicus by simply reading this book. In the immortal words of the SFAS Cadre, "Candidate, do your best."

THE 3 RULES

Always Look Cool
Never Get Lost
If You Get Lost, Look Cool

TABLE OF CONTENTS

- The Map
- Grid Reference System, Plotting Grids, and Plotting Azimuths
- Calculating Distance
- Elevation and Contours
- The Terrain Features
- Intersection and Resection
- A Note About Learning
- How To Plan A Route
- Night Navigation
- Always Look Cool
- The End Point

INTRODUCTION

Give me a decent map and a functional compass and I'll find my location anywhere in the world, no matter where I am. That's powerful. That's a life skill...*never get lost*. That's the sort of skill that builds confidence and competence. That's the sort of skill that you need to be a successful Green Beret. It's also the sort of skill that is in short supply in recent times. The introduction of GPS and the ubiquitous nature of smart phones and smart watches means that you have mapping and navigating technology at your fingertips all the time. So why would you learn to read a map and use a compass when your tech will never let you get lost? The answer is obvious...technology fails. The future of warfare is tech-enabled but it is also tech-vulnerable. Batteries drain, networks get hacked, and hardware gets exposed. Murphy gets a vote and over-reliance on this technology is a recipe for disaster.

And this is all assuming that you will always have access to your tech. If a military selection, assessment, training, or evaluation endeavor is on your horizon, you can be assured of being denied access to tech. This is especially true for those aspiring to become a Green Beret. I've spent more time than any reasonable person should spend walking around in the woods learning to navigate and now teaching others. I grew up playing outside, like normal kids should. We never had military maps, but knowing where the cool stuff was and knowing how to get there was critical kid information. Navigating through the woods between neighbors, between neighborhoods, and around our little world was essential to battling boredom. Leave the house when the sun came up, toss a couple PB&Js in a sack, and come home when the sun was setting was the order of the day. When I finally joined the Army I was well positioned to absorb all of the formal training that I got as a junior enlisted and cadet, then as a Cavalry officer, and finally into my career in special operations. With Special Forces Assessment and Selection, the Qualification course, Ranger school, Advanced Land navigation school, and the endless field training exercises, adventure races, training center rotations, and combat deployments my skills were honed. Knowing how to read a map, use a compass, and navigate competently is simply a way of life.

But let's be clear and set some expectations. This book will *NOT* teach you how to land navigate. I can sense your disappointment...you're reading a book to learn how to land navigate and the author just told you that it won't teach you how to land navigate. Keep reading, it'll all come together. This book will teach you how to read a military map. It will also describe the process of planning good routes, reading terrain from a map, navigational considerations, and the associated ancillary stuff. It will set the foundation upon which you will learn to land navigate. But in order to learn to land navigate competently you need to get out on the ground and walk. In fact, I posit that land navigation is one of the few skills that you must fail in order to appropriately learn. You must learn to get unlost. So you need to go out and get lost. This may seem contradictory given that we titled this book *Never Get Lost*, but you must understand that if you train this skillset enough you will undoubtedly get lost at some point. It is not only inevitable, but also desirable. I get lost all the time. My mind wanders, I get caught up in conversation, or fatigue simply overwhelms my cognitive capacity -- and I get lost. But I get unlost quickly. I get unlost because I remain disciplined in applying the principles outlined in this book. That's what we are doing, learning principles and setting foundations. You can't build the house unless we build this foundation first.

What is different between this book and the long list of already published manuals like ATP 3-18.14 — Special Forces Ground Mobility Operations, TC 3-21.76 — Ranger Handbook, ATP 3-50.21 — Survival, or ATP 3-20.98 — Scout Platoon? And how could we leave out FM 3-25.26 — Map Reading and Land Navigation? Well, they're manuals and they're written like manuals. They're almost punishing to read. It's like you have a robotic mid-career, starched uniform, high-and-tight haircut, knife-handing, mid-wit standing at parade-rest and lecturing from PowerPoint slides that he doesn't understand. In other words, it's suboptimal.

This book, on the other hand, is equal parts descriptive and prescriptive. It's step-by-step and it's also story-by-story. It's information and entertainment. Infotainment. You will learn the skills and you won't get punished in doing so. I will endeavor to write this the same way that I would teach it in person. A little bit informal, but very much focused and intentional. It is also comprehensive enough to get you competent without being so inclusive as to waste effort. I've navigated thousands of miles, in all conditions, across some of the most challenging terrain imaginable and I've never once needed celestial navigation. I've never had to put a stick in the ground and mark the track of the sun with a shadow to determine which direction is North. I don't look on which side of a tree the moss grows. I have a compass. I almost always have a map. So I use them to navigate. Focused.

I hear the same criticism, now used against myself, that I levied earlier...technology fails. So relying on a compass is a recipe for disaster. You have to have a Primary, Alternate, Contingency, and Emergency...but if you don't have some zero-tech, no gear, enviro-friendly method then you are doomed, right? But we have to apply some realistic logic to this exercise. You are not Special Agent Grey Man engaged in an escape and evasion event where you've been body cavity searched and stripped of your gear. Not everything has to be a worst-case scenario or a movie script. You have a compass. You just need to learn to use it. And because the central focus of this work is to prepare aspiring candidates to successfully navigate at SFAS, we know that a compass is part of the equation. So go ahead and learn how to magnetize a sewing needle on your wool sock, float it in a cup of water, and create a field expedient compass. That looks super cool and makes for a great social media post. But it doesn't move the needle (pardon the pun) when it comes to high performance land navigation. Map, compass, and skills. This is what will help you never get lost.

Basics build champions and we are going to start at the very basic basics. We are going to start slow and stay slow. I will cover everything that you <u>need</u> to know. I'll cover a whole bunch of stuff that you <u>should</u> know. And I'll regale you with nonsense that no reasonable person should ever recall...but we'll cover it anyway because it will help keep you attentive and it might make the need-to-know stuff stick a little bit better. My terminal learning objective is that you are able to take this baseline knowledge and be in a position to attend in-person live training. Maybe with me, but in-person with a competent and skilled expert who can put this stuff into practice. Someone who can safely watch you get lost and then get unlost. After some considerable time practicing these life-essential skills, you'll be in that ultimate position of achievement. You will be able to...*Never Get Lost.*

NO SHIT, THERE I WAS...THE WAR STORY

The war story is a staple of military life. The average legionnaire's existence is basically *hurry up and wait*. Everything is an emergency and requires immediate action, followed by standing around doing absolutely nothing. Sheer panic and terror, punctuated by utter boredom and idle hands. I think that this is a Universal Truth for soldiers everywhere.

This lifestyle provides for ample opportunity to tell stories to fill the time. We call them *war stories*, but they often don't involve actual war. The emphasis is on the story. A good storyteller is a valuable asset to a unit, providing much needed entertainment during dire times. A middling soldier who can spin a good yarn enjoys an elevated status amongst his peers. A good soldier who can tell good stories is worth his weight in gold.

There are some rules to war stories. They should begin with the cue to learning...something like "That reminds me of this one time...," or "True Story, listen to this...," or "My buddy told me about a guy who...." But the best attention getter is "No shit, there I was...." This is sometimes expanded to "No shit, there I was...knee deep in spent brass and hand grenade pins...." When you hear this, you can be assured of less *war* and more *story*. I prefer the shorter version.

Another rule is that it must be at least partially true, but it is more important that they convey a sentiment. The nature of war stories, like the nature of all oral history, is that they get passed down and spoken to many ears. This is especially true of war stories, and in the retelling, they sometimes get a little distorted. They are grounded in fact, but sometimes not by much. It can be difficult to separate the wheat from the chaff because so much of what a soldier is subjected to is just absurd. I think the official rule is that they must be at least 10% true at least 10% of the time.

I am going to use this time-honored storytelling tradition in this book. Some of them are mine and some of them have become mine. This is just the way that war stories work. War stories can be a helpful tool by punctuating a point or emphasizing a theme. They can also be wildly entertaining. My goal for this book is to help as many people as I can.

I promise not to take too much liberty with the rules.

1
THE MAP

"I wisely started with a map."
– J. R. R. Tolkien

The problem with teaching map reading via a book is that you need an actual map. Military maps tend to be rather large, usually about 2 x 3 feet. That's a big book. Untenable you might say. So we have a limitation that we are going to work around and I will endeavor to be as consistent as possible when referencing a map. My focus is military map reading and more specifically in preparation for successfully attending SFAS. I'd like to keep us focused on military maps and more specifically on the Camp Mackall Military Installation map, but frankly this map doesn't cover enough unique terrain to be sufficient for learning all of the terrain features. So we will reference some other maps for these particular teaching points, but we will otherwise make every effort to stay focused.

Let's start by demystifying maps and talk about what a map really is. A map is simply a two-dimensional graphic representation of an area. It is usually flat (hence two-dimensional), although they can be raised or shaped to demonstrate relief (elevation). It is usually drawn to scale as to assist in properly interpreting the area it represents. They are usually on paper, but you can certainly build terrain maps on the ground for briefings and orientation. There are specific symbols that represent specific features, to include colors and iconography. They can be hand drawn and ornamental or graphically generated and utilitarian.

A Map...Sort of.

The iconic street map kids carpet certainly qualifies as a map, even if it's unlikely to serve you well beyond the playroom. It is a two-dimensional graphic representation of an area. It has specific symbology that corresponds to items on the map; the green is grass, the grey is roads, the yellow is railroad tracks, and the buildings and locations are clearly marked. Even a child can determine locations and routes. Military maps can be similarly simple to read, but you have to know what you're looking for. Let's break down what those things are.

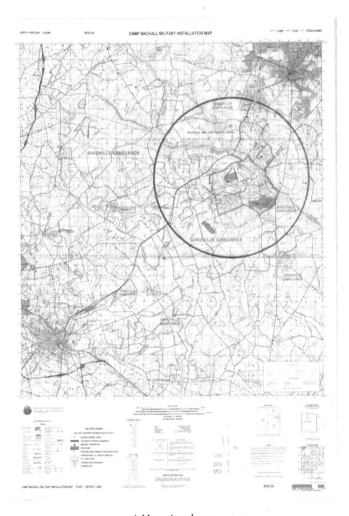

A Map...A real map.

Marginal Information

Marginal information is...wait for it...information printed in the margins of the map. Much of this information is superfluous and it's not particularly helpful for SFAS, but there are some critical elements that we will absolutely use. Let's start at the top and work our way clockwise to review just what is what.

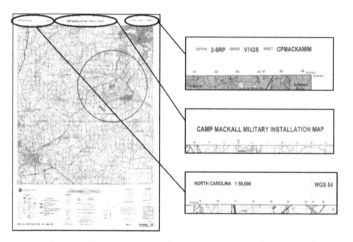

Sheet Name, Series Name, and Edition Number

Sheet Name, Series Name, and Edition Number

The top margin gives you the sheet name, the series name and scale, and the edition number. In the center is the sheet name, and this is repeated in the lower left corner as well. The sheet name is usually named after the most prominent feature, most often a town or city, in the area covered by the map. In this example the map is named Camp Mackall Military Installation Map as this sheet was specifically produced by the US Army's Sustainable Range Program (associated with the infamous *Range Control*). This is a special edition map that is not a regular part of the traditional series maps (we'll cover this in a few moments).

In the upper left corner, you will find the map series name, which usually identifies a group of similar maps that that encompass a specific area and follow the same scale and same format. You will note that this map is printed in the 1:50,000 scale which is the traditional ground navigation scale. The most common scale for US Geological Survey (USGS) maps is 1:24,000 which give you increased fidelity for detailed analysis (this is what you might use for navigating in a National or State park, for example) but is too finite for military maps because military maps must account for different maneuver forces. It's difficult to translate the few thousand meters of scale that a Soldier might walk with the scale for a mounted maneuver force on the order of ten to twenty thousand meters. Now add the twenty to thirty thousand meters you might want for artillery fires and scale up again to thirty to forty thousand meters for close aviation support. So scale becomes a balance of enough detail versus enough map coverage for all of the competing forces and after years of trial and error we seem to have settled on 1:50,000.

The upper right corner of the map marginal information depicts the edition number and is repeated in the lower left corner. The edition number usually correlates to the date of the map data, with a higher number indicating a more recent date of publication. This map was printed in 2017. If you are seeking a Camp Mackall map on your own, you should know that there are multiple versions with multiple dates, so buyer beware. I've seen for sale two "standard" versions each with three different dates (1991, 2001, and 2017) and many commercial "custom" versions of varying dates. I worked with some professional cartographers and have created a custom version that I use for the TFVooDoo Land navigation Muster events. In the grand scheme of things, none of this matters for SFAS. You will get issued whatever map you get issued and you won't have to worry about scale or series or sheet names. You won't get to choose what year you want, and it would be ill-advised to counsel the Cadre that they need to get updated maps. You fight with what you have.

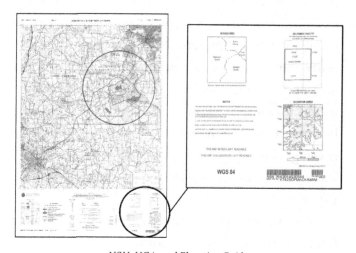

NSN, NGA, and Elevation Guide

NSN, NGA, and Elevation Guide

The lower right corner shows the National Stock Number (NSN) and National Geospatial-Intelligence Agency (NGA) reference number which is the entity that normally produces maps for the US military. You will also find the *elevation guide* (you may note the relative lack of elevation in the area), significant political *boundaries*, some administrative *notes*, and the *adjoining sheets* graphic. The *adjoining sheets* graphic shows the maps that are sequentially on the boundary of this map, and you may note as discussed earlier that this map is a special edition and overlaps four traditional map sheets. Some older versions of this map are called the Camp Mackall Military Special.

Continuing clockwise, the bottom center of the marginal information contains a bevy of useful data. In the top of this bottom section are the bar scales, which are useful in determining distances. We'll talk much more about these later; just note for now that this is where they are located on the map sheet. The bar scales also contain the contour interval, or elevation represented between various contour lines. As indicated in the elevation guide, there is not significant elevation in this area but there is enough that it will prove useful to the savvy navigator. There is some additional administrative data like projection type and datum type. Most of this data isn't really useful to you except maybe the date. Later in your operational career you might want to confirm that your entire maneuver element is operating with the same map derived from the same data, but I've never seen this be a real issue.

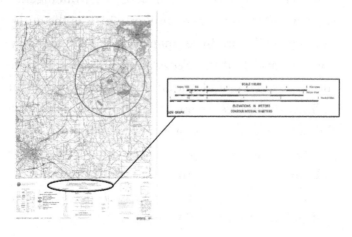

Bar Graph

Bar Graph

This center section also features the grid reference box that includes grid zone designation and some short grid plotting instructions. We will cover this information in detail in the next chapter. Ignore the slope guide and conversion graph information. You won't reference this at all for SFAS and frankly I've never referenced this data ever in my navigation endeavors. I'm certain there is some value to this data, but I've never actualized it operationally. Along with the bar scale data, the other information that you most certainly will use is the declination diagram data. Everything else you can safely ignore. You will want to be familiar with it as you advance in your navigational and mission analysis needs, but for

SFAS you really just need the map itself, the bar scales, and the declination diagram.

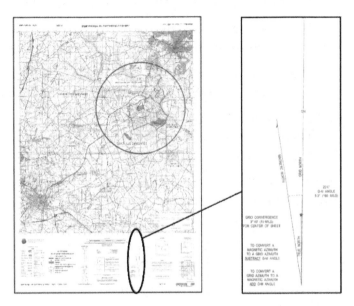

Declination Diagram

Declination Diagram

The declination diagram depicts the differences between grid north and magnetic north. There are actually three norths - true north, grid north, and magnetic north. Most people don't recognize that the magnetic poles, most often measured with a compass, are ever so slightly shifting. So we have true north that depicts the physical location of the north pole, magnetic north that depicts where the compass indicates, and grid north where the map depicts north is. This will become important when you plot directional azimuths on the map and need to translate that to a directional azimuth that you can follow with your compass. In this case the deflection, or difference between grid and magnetic direction, is a difference of 9.3 degrees. This is referred to as the Grid-Magnetic Angle, or GM Angle. For comparison, when I was testing on the STAR course in the last century (when Selection was hard, the last hard class, actually) the GM Angle was 7 degrees. And we navigated uphill. Both ways. In the snow.

There is some math associated with calculating the GM Angle and converting from grid to magnetic or magnetic to grid, but you don't need to commit this to memory because the formula is literally printed right on the map. For posterity, when converting from magnetic azimuth to grid azimuth you subtract the GM Angle. When converting from grid azimuth to magnetic azimuth you add the GM Angle. But again, it's printed right there on the map so you never have to rely on just your memory. I have never committed this formula to memory, and I have yet to need to. It is printed right on the map. We'll cover all of this azimuth stuff in later chapters, so for now just know that this is where the declination diagram is located. You can also check on local declination data by going to magnetic-declination.com and searching for the location you need.

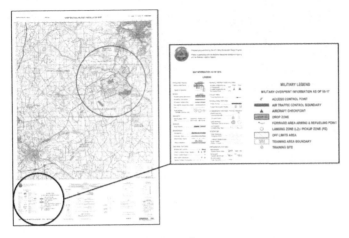

Legend

Legend

Finally, in the lower right corner of the marginal data is the legend that depicts all of the topographical features that you might find on the map. For SFAS, the only features that you will utilize are unimproved roads, trails, vegetation, and water features. There are no buildings or built-up areas to speak of. There are no improved roads (except to mark the boundaries of your maneuver area) and there are no radio towers, railroad tracks, bridges, or mines. This is one of the reasons why this is the area selected for your training; the lack of features adds to your challenge. You will become intimately familiar with the unimproved roads and trails, many of which are inconsistently annotated on your map (we'll talk techniques to account for this in later chapters). You will become uncomfortably familiar with the vegetation and water features. The hydrography -- lakes, streams, creeks, and draws with their telltale ancillary "land subject to inundation" markings (aka swamps and bogs) -- will become key navigational cues to avoid. You may also note a "Military Legend" on this example map. This is not traditionally printed on most maps but was a condition of the special printing agreement to produce this particular edition. It does not impact your SFAS navigation.

Map Colors

We'll close this marginal data analysis with a discussion of the colors that the map is printed in. Military maps are traditionally printed in five colors; red, blue, black, green, and brown (six if you include the red-brown sometimes used).

- **Red**- Used to identify cultural features, such as populated areas, main roads, and boundaries, on older maps.
- **Red-Brown**- Used to identify cultural (man-made) features such as buildings and roads, surveyed spot elevations, and labels.
- **Brown**- Used to identify relief features and elevation, such as contours on older edition maps, and cultivated land on red-light readable maps.
- **Blue**- Used to identify hydrography or water features such as lakes, swamps, rivers, and drainage.

- **Green**- Used to identify vegetation with military significance, such as woods, orchards, and vineyards.
- **Black**- Similar to red-brown, black is used to identify cultural (man-made) features such as buildings and roads, surveyed spot elevations, and all labels.

Conclusion

So that's a fairly comprehensive, but focused, overview of maps in general and military maps specifically. At SFAS, many unsuccessful Candidates have a bad habit of blaming the map for their woes. They claim that the inaccuracies (trails and unimproved roads) are too pronounced to overcome. They claim that the point they were searching for wasn't there. They will claim that the Cadre gave them poor instruction or bad guidance. They will claim that they didn't know about map cases or markers or protractors or, or, or... You see where this is going, right? Every Candidate has the same map and the same packing list. A bad carpenter blames his tools. The reality is that successful candidates prepare themselves appropriately. They follow instructions. They read the packing list. They read books on land navigation, and they seek training opportunities. Stop making excuses and start making plans. Performance is the only thing that matters.

NO SHIT, THERE I WAS...
GOOD JOB, YOU'RE FIRED.

This *war story* is a hard brag, from beginning to end. I'll admit it now. But it's good demonstration on just how important knowing how to navigate is. It's never a bad thing to know where you are. This skill was never more apparent than in Ranger School. I was lucky enough to attend Ranger School multiple times, culminating in my final attendance post Q-Course, so I was already a Green Beret. Pursuing the 'short tab' while already possessing your 'long tab' is a mixed bag. You enjoy a certain level of celebrity amongst your peers, but there is also an expectation of performance that is much higher than everyone else.

I had the bad misfortune of tripping, falling, and breaking my compass during the land navigation test in the first week of Ranger School. First and only time I've ever seen an issued lensatic compass actually break. This should have been catastrophic, but the Ranger School land navigation is fairly easy. It's a relatively small course and it's littered with points that all have the grid coordinates marked on them. I found that it was pretty simple to just walk from point to point, self-correct based on the ample way points, and I was able to finish the course without issue. When I checked in at the end, I was one of the first students done and the Ranger Instructors (RI) were mildly shocked that I did so with a smashed compass. They used this little tidbit to antagonize other students who struggled with land navigation, "Ranger Walton found all his points with a broken compass and you idiots got lost with a good one!" The performance expectation was set high now.

I happen to like patrolling and it was always a strong skill for me, so Ranger School was more about enduring the suck. I did learn quite a bit, but I found myself teaching as much as I was learning. Early on, my squad and then my platoon learned that I was a skilled navigator, reinforced by my broken compass performance, and they quickly saw the merit in asking me to plot patrol routes and then navigate the patrol to the objectives. Patrol navigation at Ranger School isn't particularly challenging, but you are well and truly smoked so even simple tasks take on a level of concentration that proves interesting. Even when my squad wasn't the lead squad in the order of march, Patrol Leaders always tasked me with being the "slack man" so that I was up front and leading us around. It kept me engaged and it's never a bad thing to be well-respected amongst your peers when peer evaluations are always on the horizon.

It became an open secret that no matter what squad was the lead squad, also designated as the security squad, that I would be task organized to that squad. As a result, I was always assigned to security, and it wasn't until late in the Mountain Phase that I even saw an objective. I was always stuck doing perimeter or ORP security. I rarely did anything during movement but navigation for the patrol. It became so obvious that the Ranger Company First Sergeant, a legendary Ranger and Delta Operator we'll call First Sergeant Tommy, decided that the company was relying on my navigation too much. Tommy was a man amongst men and would regularly walk and grade patrols, just like any junior Staff Sergeant RI, so he had both the respect of his cadre and his finger on the pulse of the students.

One morning Tommy pulled me aside and just started shooting the shit with me, like we were old drinking buddies. This is a seismic shift in the RI-to-Ranger Student dynamic. I was guarded but happy to engage and be treated like a normal human being, so we talked for about twenty minutes or so. As we were wrapping up, he pulled out a SPOT Report card. For those that have been to Ranger School, you understand the trepidation that seeing a stack of SPOT Reports might bring. If you fall asleep repeatedly, or lose a piece of gear, or have some other fatigue induced brain-fart you can quickly earn a write up and if you accumulate enough of them, you can find yourself recycled or even worse dropped from training. You might also find yourself earning a positive report. But these are so rare that the overwhelming feeling upon seeing a card pulled out is of dread. So Tommy hands me this card that is already filled out, and it's a positive one. It details my performance thus far and highlights my excellent navigation skills. It notes that I have led the patrol to nearly every objective without any deviation and that I am embodying the Ranger ideals. He goes over it and has me sign it. I'm a little stunned.

First Sergeant Tommy then calls the entire patrol into the center to address us. He had this very cool habit of bringing us all in whenever he walked with the patrol to just impart words of wisdom and mentor us a bit. He would also quietly, almost stoically, lead us in a recitation of the Ranger Creed. Whenever you recite the Ranger Creed in Ranger School it becomes this sort of rote screaming match where your worthiness of the words is measured by the volume of your voice, not the manifestation of any credence. It just meant more when Tommy did it quietly with us. He closed out this mentorship session by reading my positive SPOT report to the entire patrol. It was a glowing report and I'm not embarrassed to admit that I was enjoying the accolades.

He finished his reading by announcing that, "Ranger Walton is no longer permitted to lead this patrol in any manner. He will not plan, advise, or assist in any navigation tasks. In fact, Ranger Walton is not allowed to be in the lead squad during movement or be within 100 meters of the patrol leadership. This patrol will learn to navigate on their own. Good job Ranger Walton, you're fired." I was off the hook. I got to return to my 'junior rifleman number 2' role and dedicate all of my efforts to just staying awake.

When I got to my ODA a few months later I was recounting this tale to my new teammates, including my 18F who had been a Mountain RI and had served in the Ranger Regiment. Stew, my 18F, was a bit of a hard ass and had a healthy bit of disdain for the Officer class. He was slow to praise and naturally protective of the Ranger brand. I could see his reservation in accepting my story. He confirmed who the First Sergeant was and let it go. I saw him step outside soon after and he returned about ten minutes later. He walked up to me and quietly just shook my hand and said, "Welcome to the Team."

Turns out that Stew and Tommy were old buddies. Tommy was one of Stew's Squad Leaders in the Ranger Regiment and later they served together as RIs in the Mountains. Stew, being naturally distrustful of officer performance, called Tommy to validate my story. Tommy was more than happy to sing my praises and for Stew, that was like being kissed into the Mafia. I was a made man now. Stew went on to become my Team Sergeant and we built a high-performing ODA that was the envy of many. I made certain to remind Stew about that SPOT report whenever he inevitably questioned my intellect..." Stew, do I need to call Tommy?"

Your reputation in SOF is usually well established before your shadow ever darkens the doorstep of your next assignment. But it is never more than a phone call away.

2

GRID REFERENCE SYSTEM, PLOTTING GRIDS, AND PLOTTING AZIMUTH

"Bring a Compass. It's Awkward When You Have to Eat Your Friends"

The world is big (no shit). If someone asked you where you were from you wouldn't say, "I'm from Earth." Because that would sound dumb, but also because it wouldn't really narrow down where you're from. You could say, "I'm from North America." You would still sound dumb, but at least we would start to narrow down where you are from. You could further refine where you are from by identifying the country, state, city, and even neighborhood. "I'm from the US. I live in New York City, New York. The Bronx, actually." We would probably already know this information because you would have that horribly grating Bronx accent and attitude, but you understand how this descriptive system of country, state, city, neighborhood is a formal method of narrowing down a location. But it's irregular. It doesn't really conform to a repeatable, logical, predictable system. These political boundaries are subject to odd eccentricities and peculiarities. They sometimes follow geographic features, like rivers, but they often don't. These boundaries sometimes follow specific latitude and longitudes, but they often meander randomly. And the naming conventions are wholly unpredictable. North Carolina and South Carolina seem logical, but why isn't Louisiana just called East Texas? So while its helpful, it's not finite. In order to gain some fidelity it is best to establish a repeatable, predictable, logical, and calibrated system, like a grid system that you could reference.

A grid reference system, any grid reference system, allows the navigator to locate a specific position on the map by plotting an X (left and right) and Y (top and bottom) axis. It is a spatial reference tool that can pinpoint the referenced location. Many novice map readers struggle with the system, especially when it involves 8 digits. So let's start this with a very basic primer to help understand the concept, then build to the actual plotting. Let's take a simple box and divide it into 4 sections, or quadrants.

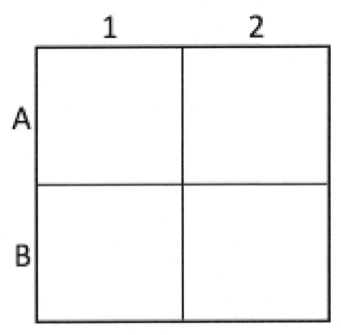

A simple grid reference system

In this example there are 4 squares, or 4 "grid squares." The upper left, upper right. Lower left, and lower right. We have assigned numbers to the X axis (left and right) and letters to the Y axis (top and bottom or up and down). In establishing these boxes and overlaying these numbers and letters, we have developed a way to identify the location of an object on this "map." For example:

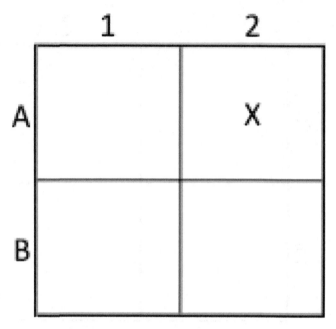

A point...on a simple grid reference system

The X (X marks the spot, right?) on this "map" is located in the 2A quadrant. The traditional way, and the way that the military plots locations, is to read the location designator left and right first, and then up and down. Remember the phrase, "Read right, then up"; I've been plotting points on a military map for over three decades and I still say, "Read right, then up" every single time that I plot a point. Sometimes I say it in my head, but sometimes I go full blown weirdo and say it aloud. But I still say, "Read right, then up." Every single time, and I never plot my coordinates incorrectly. There is a lesson in that consistency. This will prove helpful when we get to a Military Grid Reference System and plotting points. Back to our simple "map." We have the X in grid square 2A. Simple. But it's a small map, and it's a simple read. What if we made that map bigger? Maybe twice as big?!?

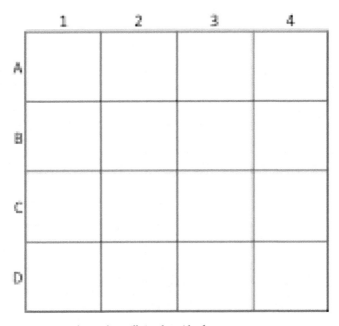

A larger, but still simple, grid reference system

Now, there are four positions on the X axis – 1, 2, 3, and 4 – and there are four positions on the Y axis – A, B, C, and D. But the process is exactly the same. So we can extend the same logic and locate a position on this "map," correct?

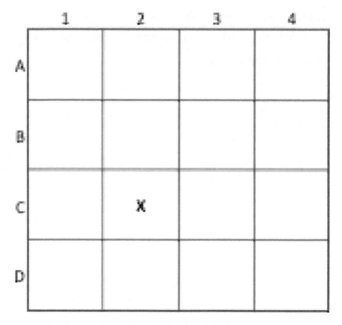

A point on a larger, but still simple, grid reference system

The X, the new designated location, is now located in the 2C (read right, then up) grid square. This is fairly simple and you likely aren't struggling to understand this concept. But we just spent the last chapter talking about all of the colors, features, markers, and information that are usually on a map. And that stuff muddies the water a bit. Or does it?

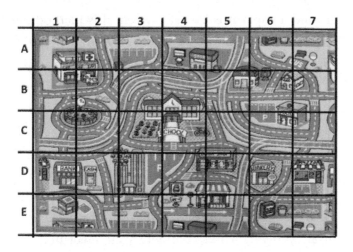

A map on a larger, but still simple, grid reference system

Now we overlay our grid onto a "map" and there are bunch of extra things to orient ourselves to. We have an expanded grid system on the X and Y axis grid, we have buildings and roads, and we have all sorts of features that "muddy" our picture. But the concepts are exactly the same. Read right, then up. On our new expanded map, the Bank is located (mostly) in 1D and if I asked you what building was located in 6D you could pretty easily plot, using the X and Y axis, that the Bakery was located in 6D. Thus, we can start to see that the concept of a grid reference system is transferable. As long as I have a standardized grid system overlayed on a standardized image, I can easily, consistently, and most important precisely locate and communicate locations using this process.

So far, we have been using an arbitrary grid identification system with simple numbers and simple letters (1,2,3 and A, B, C), and because we haven't created a super precise grid system this method has served us well. But you could see how there might be some limits to this simple grid system. For example, if I asked you where the school was located, could you use our system and identify its location? On our map the school occupies several grid squares; 3B, 4B, 3C, and 4C. If you had to give precise directions to someone to have them meet you at the school, there is some room for interpretation. If they were waiting for you at 3B, but you were actually in the parking lot in 4C you might not complete your meet up. Now imagine a smaller more detailed scale, more concealing features, and the potential for someone shooting at you. You would want to be more precise.

As such, military maps must allow for this additional and unique precision. They follow a common and transferable grid system (called the Military Grid Reference System or MGRS), they have a common color scheme, a common scale, and common features, just like we reviewed in chapter 1. The military is often criticized for being too strict on its "lowest common denominator" mentality and many of those criticisms are valid. But when it comes to maps, you want it to be precise, repeatable, and accurate. And drawing back from my earlier pedantry of "read right, then up" we should review this MGRS and how it works.

What the Hell is Lat Long?

Latitude and Longitude are simply a standardized grid system that measures a location north or south of the Equator (latitude), while longitude is a measurement of location east or west of the Prime Meridian. The Prime Meridian is an imaginary north-south line that passes through both geographic poles and Greenwich, England. It's just another grid system. It is an ancient construct having been originally theorized by a Greek mathematician and philosopher and was further developed by another Greek mathematician in the 3rd Century BC. Much of that work was predicated on the ancient Phoenician method that was based on astronomical observations. However, Longitude cannot be determined via astronomical observations so the problem of accurately determining longitude from any location on Earth continued up until 1762, when English inventor John Harrison developed a method for determining longitude based upon highly precise timekeeping. This is why lat/long coordinates include both degrees and minutes-seconds. None of this is important for your navigation at SFAS except to put it in context. Old dudes in robes figured out how to circumnavigate the globe with sticks and stars, an old dude with a wig added timekeeping, so your map and compass are certainly sufficient if you can learn how to use them. No robes or wigs required.

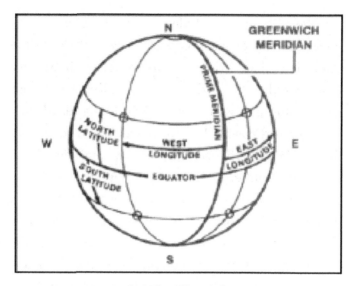

Latitude and Longitude

Military Grid Reference System (MGRS)

Before we dive into the Military Grid Reference System, we should set the stage with the Universal Transverse Mercator System (UTM), from which the MGRS is derived. Lat/long uses degrees-minutes-seconds and UTM divides the Earth into 60 zones, each 6° of longitude in width (excluding the polar regions which use the universal polar stereographic (UPS) coordinate system). As discussed in chapter 1, the UTM system plays a role in MGRS with zones (lower center on the marginal data). In a moment we'll talk about 4, 6, and 8 digit grid locations, but these grids, absent of the UTM zone could be plotted in multiple locations across the globe. It's sort of like how multiple cities of the identical name might be confused without a state designator. Springfield, Massachusetts or Springfield, Illinois? So the UTM is a bit like a state designation. The UTM zone refines that location to specific UTM zones based on the UTM grid system and for SFAS the UTM zone is 17S PU. Every grid that you must plot at SFAS will have the 17S PU designator so you won't need to distinguish it separately, just know that MGRS grids can have this characteristic identifier.

So back to MGRS. The UTM zone system identifies a zone (17S) and further divides into a 100,000 x 100,000 meter grid square and identifies a unique two-letter code, in the case of Camp Mackall that code is PU. Thus, every point that you plot is prefaced by 17S PU (in military parlance, Seventeen Sierra, Papa Uniform). It is technically correct to use this prefix whenever you are communicating a grid location, but brevity and common sense often omit them and you simply communicate the numbered grid location, which we are building to. But by using the UTM and its unique numbering system we can already narrow down our location to a 100,000 x 100,000 meter square which is pretty good, considering the grand scale we are working with. And we can find this information in the marginal data on the map, in the lower center portion.

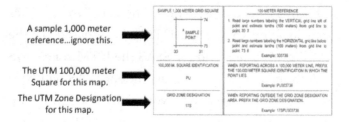

From the marginal map data

BUT WE WANT to avoid walking around a massive 100,000 x 100,000 meter square, so we have a system to do that. As described in Chapter 1, our maps have a numbered grid overlayed on them. Sometimes the numbers are highlighted, like in our example, but they are often not. You can use your own highlighter to make finding them easier and later in the advanced tips section you will want to note their location when determining how to fold your maps. These numbers on the map now identify 1,000 x 1,000 meter squares. Now we're getting somewhere, but 1,000 by 1,000 is still pretty large and we want to be as precise as possible, so we need another tool to continue to refine this grid we have established.

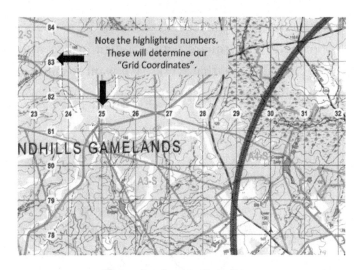

The numbers for our grid coordinates

Note the highlighted numbers. These are the numbers from which we will determine our "grid coordinates." Every location on this map is already prefixed by the 17S PU, and now we see where the numbers come from. But what do they mean? The 17S PU gets us, in an orderly, predictable, and repeatable manner down to a 100,000 x 100,000 meter square, and the numbers now help us get to 1,000 x 1000 (a 4-digit grid) resolution, 100 x 100 (a 6-digit grid) resolution, and eventually 10 x 10 meter square (and 8-digit grid) resolution. You can use a 10-digit grid to get to a 1 x 1 meter square, but that's really not relevant for land navigation. This sort of precision is impossible with your standard tools. A mechanical pencil is likely to make a mark on a map exceeding even a 10 x 10 level of resolution, so a 1 x 1 is simply not tenable. If we were laser designating targets for precision fires then yes, but not for land navigation.

So following this orderly, predictable, and repeatable method we can calculate, down to a 1,000 x 1,000 meter grid square with a 4-digit grid coordinate. Let's see how that works. We want to plot the location of that 6 way intersection of unimproved roads that is circled here.

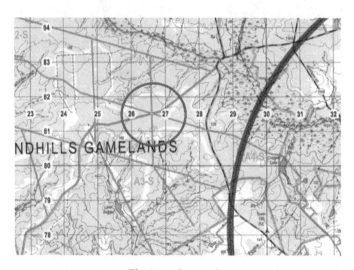

The 6-way Intersection

We know that every location on this map is prefixed by 17S PU, and we can see the highlighted numbers, but how do we read and interpret them? Remember our mantra of "Read right, then up?" Let's put that into practice. Let's read the left to right numbers and determine which box corresponds to the location we are finding. From left to right we see them denoted as 23, 24, 25, 26, 27, 28, 29, etc. We can see that the road intersection is located in the left to right box as box 26. Because we are reading left to right (**read right**, then up) we are identifying the lower left corner of the intersecting grid lines. So we start with 26...our 4-digit grid coordinate then looks like 17S PU 26??. So, let's find the two missing ?? numbers next. We read right, then up. Reading up we see the numbers denoted as 78, 79, 80, 81, 82, 83, etc. We can see that the road intersection in the bottom to top box as 81. Because we are reading bottom to top (read right, **then up**) we are identifying the lower left corner of the intersecting grid lines. So we started with 26 as our left to right number and in a 4-digit grid coordinate, we add 81, and our grid coordinate then looks like 17S PU 2681. Read right – 26, then up – 81. Not bad.

This calculation now tells us, with a resolution of 1,000 x 1,000 meters, what grid box our intersection is in. 2681. More correct 17S PU 2681. And because the numbered grid lines extend across the entire map sheet in an orderly, predictable, and repeatable manner we can locate any feature on the map within a 1,000 meter fidelity with a simple 4-digit grid coordinate. But we are walking, with a heavy ruck, so we want to avoid wandering around this 1,000-meter box looking for our road intersection. If we were communicating to another person that we wanted them to meet us at the road intersection in grid 2681, they might need more finite directions. What if there were two intersections in that same grid box? What if there was no terrain feature and we were describing a location where we wanted an airplane to drop a resupply bundle? We need more fidelity. We need something to continue this orderly, predictable, and repeatable methodology. We need more grid.

Plotting a Grid

Plotting a grid is a little bit science and a little bit art. This is where the *protractor* comes into the equation. The protractor, or Graphic Training Aid (GTA) 05-02-012 is officially called the Coordinate Scale and Protractor. There are many commercial variants, of all scales, but they mostly look very similar to the GTA shown here, if not identical except for branding. Only weirdoes will call it anything other than a "protractor." The protractor is a clear plastic sheet about 5 inches by 5 inches square that serves to both refine the grid system and determine azimuth, or magnetic direction (we'll cover this later). You will note that the protractor has three coordinate scales on it (the triangles), 1/25,000 (the big triangle), 1/100,000 (the little triangle), and 1/50,000. These scales coincide with the common scales that maps are printed in. Let's refer back to Chapter 1 and determine our scale by looking at the bottom center and upper left of the marginal information what scale our map is.

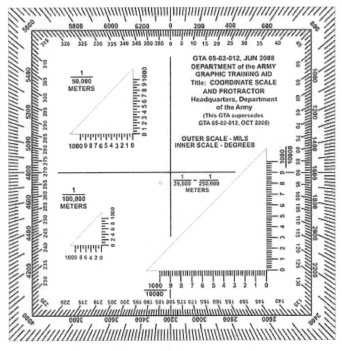

The protractor, or Graphic Training Aid (GTA) 05-02-012 is officially called the Coordinate Scale and Protractor

We can see that our map (as is almost always the case for military maps) is 1:50,000. This is often spoken as, "One over fifty", but is understood as 1 over 50,000. This is the exclusive scale that you will use in SFAS. I have used many other scale maps (and protractors) in advanced training, but 1:50K is the most common by a massive margin. If you have significant orienteering or outdoor experience you are likely familiar with the 1:24,000 USGS (US Geological Survey) maps which are in common use. But now that we know we are using 1:50,000 we can go back to the protractor and note that the triangle in the upper left corner is the one we will use for this map. You will note that this triangle is numbered...in an orderly, predictable, and repeatable manner. Those numbers help us form another grid.

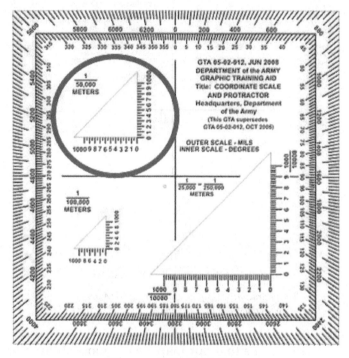

The 1:50,000 Coordinate Scale

Using the 1:50,000 coordinate scale triangle (upper left) we see that it is divided into ten more sections, on both the horizontal and vertical axis. You have to use a touch of imagination here, but its not much of a leap to visualize this grid. These orderly, predictable, and repeatable numbers help to establish another smaller scale grid. In order to determine where in that square we are locating we need more numbers, or more grid. We can visualize this as such.

Never Get Lost 53

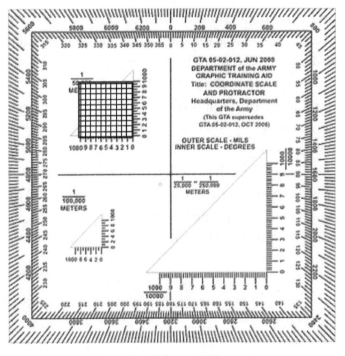

More grid for your grid...

This new smaller scaled grid fills the entire 1,000 x 1,000 meter grid box and turns it into 100 different 100 x 100 meter grid boxes. So we can now overlay that new smaller scale grid box over the map and by using the same orderly, predictable, and repeatable process we can further locate a position within the grid. But...the military is nothing if not complex. This next measurement is a little bit different than the 4-digit grid where the numbers are logically sequential. Now, even though we have the new smaller grid, we count the numbers differently. Because the new smaller grid is imaginary, we want to ensure more accuracy rather than just estimating position.

The 4-digit grid gives us the 1000 meter grid and we take the protractor and determine the more finite 6-digit (100 meter fidelity) and 8-digit (10 meter fidelity) grids as such. For demonstration and clarity purposes I have removed the bulk of the protractor and simply kept the 1:50,000 scales from the upper left portion. In the following exercises you will only see these scales, just understand that normally you will have the rest of the protractor, obviously. We'll talk more of tips and trick for maximizing your protractor, and other gear, in later chapters.

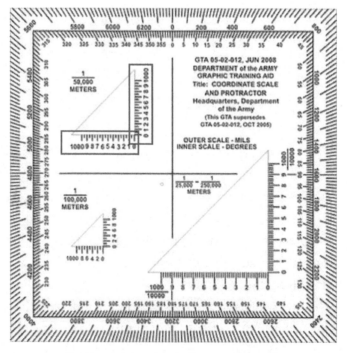

For demonstration purposes...prepare to be wowed.

As such, overlayed on the map it would look as such:

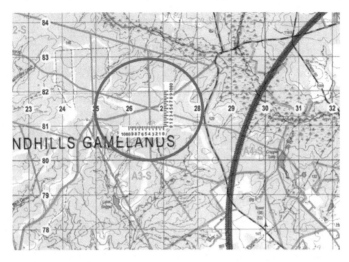

...be wowed. Amazing!

If we were to overlay our imaginary 100 x 100 meter grid it might look like this:

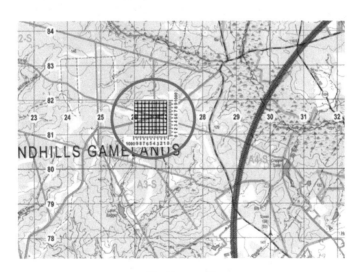

Double wowed!

But again, because we are reading right, then up we need to align our protractor to read correctly by sliding the protractor to the left so that the horizontal scales index to the position we are reading, in this case the 6-way intersection. We are still reading from the 26/81 intersecting lines in the lower left. Your protractor will now look as such:

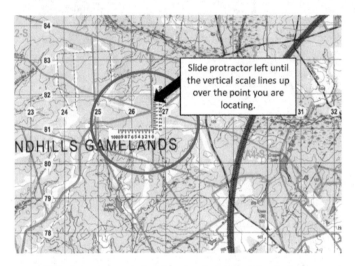

Slide the protractor to the left so that the horizontal scales index to the position we are reading.

Now, the protractor is properly aligned both horizontally and vertically with the target (the 6-way intersection) and the grid lines. In this position we can now read the next 2 numbers in our 6-digit (100 meter fidelity) grid. We can now go from 17S PU 26?81? to 17S PU 266815.

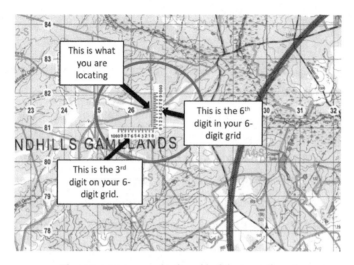

The protractor is properly aligned both horizontally and vertically with the target (the 6-way intersection) and the grid lines.

That gets us a 6-digit grid that gets to 100 meter fidelity, but we want a little bit more accuracy. We want to get down to 10 meters fidelity and this means getting an 8-digit grid. If you are looking at the 1:50,000 coordinate scale you may note that we are getting pretty fine here. The 1000 x 1000 grid on a 1:50,000 map is only about an inch by an inch. So a 6-digit grid is slicing up that 1 inch by increments of 10. That's $1/10^{th}$ of an inch. Accuracy is important but we're already slicing this onion really thin here. So I want you to keep this in mind as we get down to an 8-digit grid. There is a little suspension of disbelief involved. Some interpretation. Some calculation. Some estimation, or maybe better "guess-timation."

A quick analysis of the map shows us that an improved road (denoted by the red and white striped lines), when drawn on the map, looks to be about 50 meters wide. That's pretty large and not entirely accurate. So if you were trying to calculate an 8-digit grid (with 10 meters of fidelity) and you were using an improved road drawn to 50 meters fidelity as your benchmark, you can see how there is a little room for interpretation. That's not to say that you shouldn't trust your map or your calculations, just understand that there are some limitations that you need to account for. So that $1/10^{th}$ of inch measurement from the coordinate scale will need to be further delineated another 10 times in order to get us to an 8-digit grid.

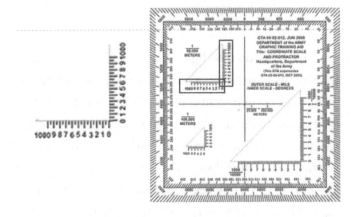

Get ready for some "guess-timation"

You are going to take this scale that ends up being about an inch and to get to an 8-digit grid you must further divide each increment into another 10. This task is made slightly easier in that there are little tick marks between each of the numbers. If you look closely, you can see the little half sized tick marks that denote the midway point between numbers which represent the 5 position. So now, with great faith in your ability to both accurately locate the point (are you measuring the east side of the road intersection or the west), correctly position the protractor (are you looking down at it from exactly overhead), and properly read the map (is the map laid flat and true) you will determine the last digits. We have 266815 and we need to get to 266?815?. As such, my best estimation is something like 17S PU 26618151. This 8-digit grid coordinate now gives us a precise location, down to 10 meters of fidelity. This same process is used for both determining the grid coordinate of a position marked on the map and the reverse of being given a grid coordinate and marking its position on the map.

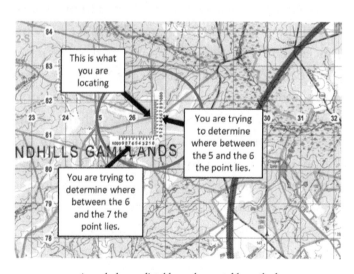

An orderly, predictable, and repeatable method

You may recall that we described plotting a point as a little bit art and a little bit science. The science part is that we can fairly precisely plot the 6-digit grid. It follows an orderly, predictable, and repeatable method. But when we are getting down to an 8-digit grid the little potential vagaries like angle of viewing, the thickness of the line, or the steadiness of the map can shape the artfulness of this last little measurement. Did you slide the protractor a little bit? Did you mark the location with a .5mm or a .7mm mechanical pencil or (heaven help you) a marker? It's still an orderly, predictable, and repeatable method, but I find that this last little calculation is so subject to these little vagaries that for most applications a 6-digit grid coordinate will suffice. But this is heresy in the land navigation world. No self-respecting instructor would espouse this position. More accurate is always better. More precision is critical to not getting lost. Wring every margin of error out of the process that you possibly can. This is my official stance. My unofficial stance is that I've never once been able to walk an extended route, contend with all of the quirks that I introduce into my route with my idiosyncrasies, and contend with the terrain – the real-world terrain that I am traversing – for this sort of accuracy to really matter. Officially unofficial. That's my stance. Learn to plot an 8-digit grid and learn to navigate on the real terrain. Art and science. Like a proper modern-day Renaissance Man.

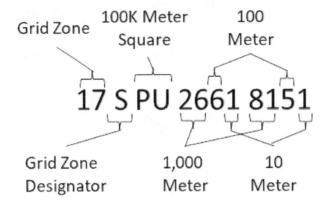

There is such a thing as a ten-digit grid that gets us down to 1 meter, but that's not relevant to SFAS and other than precision fires you are unlikely to ever use a ten-digit grid. Otherwise, this is how you plot a grid coordinate. An orderly, predictable, and repeatable process.

Plotting an Azimuth

Since we have our protractor out already, let's review how to plot an azimuth. On the perimeter of the protractor, you will find two sets of numbers. The outside set of numbers are mils; you will note that it is calibrated to 6400 mils. That's a lot of delineation that produces high precision measurements. Logically, mils are used for artillery fires where high precision is required. But it's too precise for land navigation, if such a condition exists. The reality is that you will struggle to hold a precision azimuth with just 360 degrees of delineation, so 6400 is just overkill. That's the outer ring on the protractor. The inner ring is degrees, and as previously indicated is delineated to 360. A standard measurement that produces an orderly, predictable, and repeatable calculation. You can forget the mils and we'll talk about some tips and tricks later in the book about how to prepare you protractor for maximized utility.

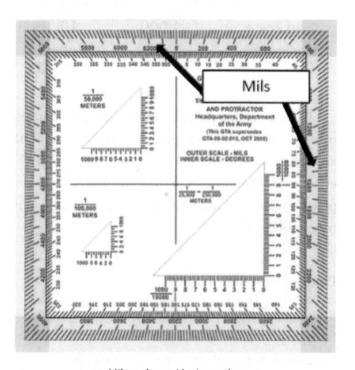

Mils on the outside...ignore them

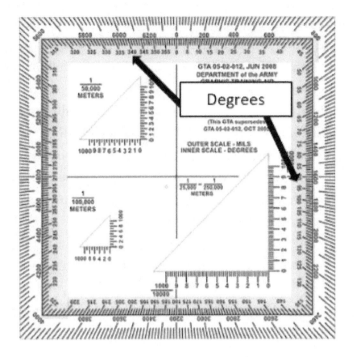

Degrees on the inside...don't ignore them

In the center of the protractor you will find a crosshair. This is a critical component that is absolutely essential to understanding how to accurately plot an azimuth. I find that when evaluating students that struggle with plotting accurate azimuths it is either misaligning the crosshairs or not extending the tracking line long enough (which we'll get to in a moment). You may have mistakenly gotten the impression that I am an advocate for the "Good Enough" school of thought here. I am not. I am in some very select instances, but by and large I am very much pro-precision. I have often said that perfection is the enemy of good enough and good enough is often all that a situation requires, but these are heuristics (mental shortcuts) that are best left to the more experienced operators. When you are just learning a new skill, you should always default to more precision is better. When you can learn to inherently read terrain, visually calculate distances, and innately understand route selection then you can defer to good enough, but not yet. So you have to be very precise when using these crosshairs. You must be precise in centering them directly over your start point AND you must be very precise in positioning the crosshairs so that they are exactly aligned north/south and east/west. Conveniently, the map is riddled with grid squares that easily allows you to do this. But take your time and be mindful of the importance of this accuracy.

Never Get Lost   65

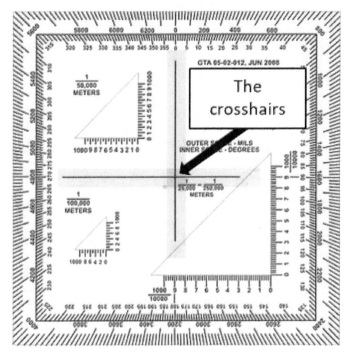

The crosshairs...aim small, miss small

In the following graphics, we are going to use a visually modified protractor to make the process a touch clearer. We have removed the mils scale on the outside perimeter, we have removed the ancillary data (like the coordinate scales), and we have enhanced the crosshairs in red to highlight their location. This will allow us to see more of the map, reduce visual noise, and hopefully temper a little confusion. The protractor is functionally identical, just visually modified for demonstration purposes. Our demonstration protractor looks as such:

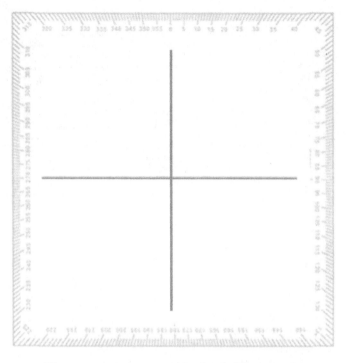

We spare no expense...a custom, hand-crafted, demonstration protractor

Before we start, we should briefly address the difference between grid azimuth and magnetic azimuth. We're simply going to mostly reprint the passage from earlier in Chapter 1 and add some amplifying narrative.

There are actually three norths...true north, grid north, and magnetic north. Don't worry about true north. This isn't some episode of *Naked and Afraid* where you have to align your spirit energy and open your third eye to get the Harvest Moon to reveal its secrets. Every time I've attended some sort of "advanced land navigation training" they make great pains to talk about *true north*, yet I've never once had to reference true north in the wild whilst navigating. I'm certain that it means something to someone, like a cartographer or something. But for the average navigator it's useless trivia. I only mention it here because in a book about land navigation it seems anathema not to. But again, I've never once had this topic come up in practicum. But grid north and magnetic north are important.

The wandering pole

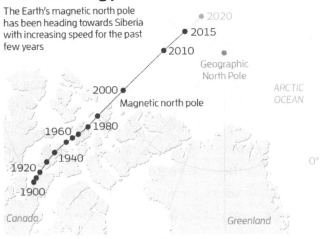

North wanders more than a lost Candidate

Most people don't recognize that the magnetic poles, most often measured with a compass, are ever so slightly shifting. So we have true north that depicts the physical location of the north pole, magnetic north that depicts where the compass indicates, and grid north where the map depicts north is. The map differs from the magnetic because the map is a flat representation of the round earth. The further you move from the equator, when the roundness is minimized, the more extreme the difference is. This will become important when you plot directional azimuths on the map and need to translate that to a directional azimuth that you can follow with your compass. This is called the deflection, or difference between grid and magnetic direction. This is also referred to as the Grid-Magnetic Angle, or GM Angle. The declination diagram, in the lower margin of the map, depicts the differences between grid north and magnetic north and tells you the formula, to add or subtract, to complete those calculations.

Yes, there is some math associated with calculating the GM Angle and converting from grid to magnetic or magnetic to grid, but you don't need to commit this to memory because the formula is literally printed right on the map. For posterity, when converting from magnetic azimuth to grid azimuth you subtract the GM Angle. When converting from grid azimuth to magnetic azimuth you add the GM Angle. But again, it's printed right there on the map so you never have to rely on just your memory. I have never committed this formula to memory, and I have yet to need to. It is printed right on the map. You can also check on local declination data by going to magnetic-declination.com and searching for the location you need. Now back to plotting azimuths on the map.

The first step in plotting an azimuth is to determine the two points, a start point and an end point for example, between which you want to determine the direction. Let's pick two points on the map to start this process; we'll go with the 6-way intersection in the 2681 grid square, or more appropriately the 6-way intersection in the 17SPU2681 grid square [You start to see here how the pedantic accuracy and strict protocols can sometimes get in the way? It is absolutely correct but saying the entire phrase of "Seventeen Sierra Papa Uniform Two Six Eight One" is a mouthful compared to "Two Six Eight One." One is correct, one is not. But one is bothersome and one is brief. We usually defer to brevity. It's one of the military paradoxes that good operators learn to navigate around, no pun intended.] Back to the map, we'll start at the 6-way intersection and our second point will be the southern end of Lake Bagget in the 2575 grid square, right where the little creek exits the lake. Start point and end point.

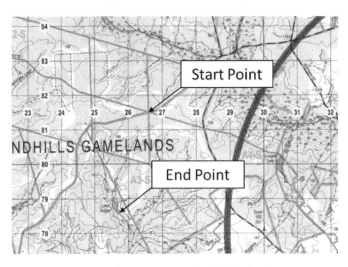

Start at the intersection and end at lake/creek

Having established these two points, we now draw a line connecting them. But we want to mindful that we are not just connecting the dots, we are drawing this line to take a measurement and that measurement is on the perimeter of the protractor. So this line must be long enough that it extends to the point that it can be seen on this perimeter, as such:

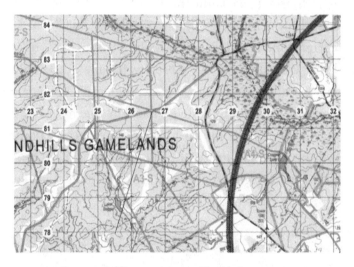

Our extended line, so we can read it with the protractor

A note about accuracy: I know that we just discussed the silly pedantry of "too much accuracy," but I also noted that it was a paradox. So let's get into it again. In the above example I didn't actually place a dot on the map to indicate my start point and my end point. But if I did place a dot and I had the temerity to use a fat-tipped marker or a Marine Corps Marking Device (aka crayon), the accuracy of my pending azimuth calculation might be subject to the eccentricities of that marking. Do I take my measurement from the center of the dot or the side of the dot? Is the dot so large that it obscures my true location? What about my line? Is my line so thick that it covers several degrees of direction? What about if I combine a big fat dot and a thicc (with 2 C's!) line to generate some real margin for error? Combine this with an odd viewing angle, a wonky map placement, and a little sleep deprivation and you can see where this cascading error margin can give you some potentially imprecise measurements. As such, I always recommend a .5mm mechanical pencil and allowing yourself a few extra moments to mark these positions carefully. Just get in the habit of these little techniques now and make them something that you always do and never have to worry about. Be accurate when you can, so that you can manage the ramifications of inaccuracy more easily later. A visual here to make my point...this is an 8-degree difference.

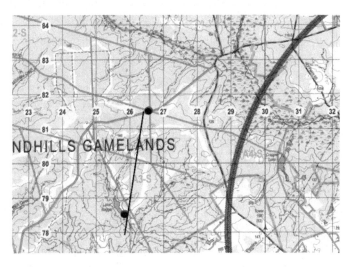

Aim small, miss small. Accuracy matters and little stuff adds up.

So now I that I have my start point (the intersection), my end point (the southern tip of Lake Bagget) and my intended azimuth (the line that I drew between the two points...and that I extended a little bit so that I could read it with my protractor). The next step is to measure the direction that this line is pointing. This is known as the azimuth. To do this I place my protractor on the map and align those crosshairs directly over my start point. Then, being careful to keep my protractor centered, I align it in the proper direction. Obviously the 0/360 (a circle is 360 degrees) is to the north or top of the map. You can use the east/west and north/south grid lines and the crosshair lines to ensure proper alignment. Keep them completely parallel. It takes a few practiced tries, but there is nothing advanced about this technique. It is made easier by laying the map on a flat consistent surface, viewing the map from directly overhead, and taking our time and staying relaxed but focused.

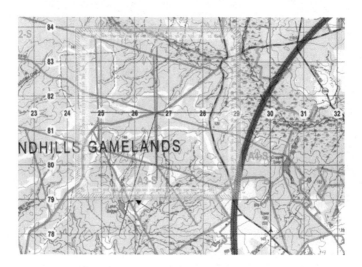

Take your time and line everything up precisely.

This graphic looks pretty simple, and it is, but let's review what's going on here so you don't miss anything. The start point (the intersection) and the end point (Lake Bagget) are very accurately plotted. The directional azimuth line is very carefully drawn to connect the precise start and end points and it is drawn with a very thin (aka precise) line. We always recommend a .5mm pencil, so you can erase it later. The red crosshairs (remember, these are only red on our "demonstration" protractor) are centered directly over our start point. Just a few smidges (a technical term that I just made up) off to the flank and your measurement will be off. The protractor itself is precisely aligned. This is most easily done by aligning the crosshairs with the north/south lines, in this case the 26 and 27 lines. You're not overlaying them, as this would de-center your crosshairs. You are simply aligning them to ensure they are running in exact parallel. You can further check your alignment by using the crosshairs and the east/west lines, in this case the 81 and 82 lines.

What you end up with is a precise measurement. Because you took the time to carefully and precisely plot the individual elements - the start point, end point, and azimuth line – AND you took the time to carefully align the protractor – centered over the start point and aligned east/west and north/south, you can now simply look to where the azimuth line that you drew passes under the perimeter of the protractor and determine the proper azimuth. In this case you can see that the grid azimuth is 199 degrees. There is no secret to this technique, it is simply the technique. This is how you plot an azimuth. Slow, deliberate, precise procedures executed with discipline, and you will get accurate measurements. I always replot every azimuth twice, just to make sure that I didn't mess up any steps or introduce any error. When aspiring navigators screw this process up its usually something obvious. It is not uncommon to plot your start or end points in an adjacent grid square by misreading the coordinates or rushing the plotting. It is not uncommon to mistakenly place your protractor on the end point instead of the start point. And it is quite common to misalign the protractor a few degrees by not paying attention to the alignment cues that we provided above. A slight twist in the protractor can throw you off several degrees. We'll cover some "advanced tactics, techniques, and procedures" later in the book that can help manage some this induced error, but for now just concentrate on deliberate and precise execution of the fundamentals as we have discussed.

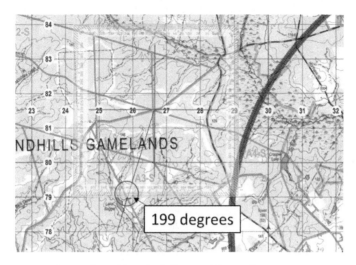

A precision process for a precision reading.

If you were navigating directly from the start point to the end point, a technique called 'dead reckoning,' then you now have half of your required information. You need the distance and the direction and now that you have the azimuth (you, of course, need to still convert this to a magnetic azimuth) then you next need to calculate the distance. We'll cover this topic in depth (distance, direction, and how to manage them appropriately) when we move from simple map reading to actual land navigation. But we aren't there yet. For now, you have a good basic understanding of what a map is and what the various sundry markings mean. You understand the grid reference system and how that system helps you plot a grid coordinate (an orderly, predictable, and repeatable process). Finally, you now know how to plot an azimuth, a grid azimuth (we'll get to conversions and compasses later). You are laying down a foundation of knowledge that we will roll together into a skill as we advance. Hang in there.

LARS

There exists in the land navigation world an insistence that we complicate things to the extreme. At many military schools they add superfluous rules to make things as stressful as possible. Many schools add a rucksack and make the routes as long as possible. Go heavy or go home! No flashlights unless you're under a poncho. No handrailing roads. Don't do this, and you can't do that. SFAS represents the pinnacle of this methodology. I understand the intent, and I support it in most cases. But I still find it bothersome. Especially in the extreme. The *Left Add, Right Subtract* mnemonic is squarely in this camp.

Also known as LARS, this prompt exists to remind us how to convert azimuths. Let's be clear – the only reason that I'm covering this is because I am a product of the system. Like a young juvenile offender that has spent his entire formative years in prison, I am indoctrinated in the military land navigation system. And the system demands that I cover certain topics. LARS is one of them. I've navigated damn near the entire globe from deserts to jungles, from mountains to plains, from rainforests to grasslands. I've used issued maps, road maps, tourist maps, and hand drawn maps. Day and night, mounted, dismounted, and flying. I'm not *THE* expert, but I am certainly *AN* expert. I've never once had to use this mnemonic. But the guards are changing shift, so I'm standing in my cell door to tell you about LARS.

The way it works is that when we are converting from grid to magnetic, magnetic to grid, using a westerly or an easterly angle all that we must remember is that when we are moving left on the declination diagram we add the G-M angle, and when we are moving right on the declination diagram we subtract the G-M angle. Left add, right subtract. LARS. There is a 99% chance that you will never use this in your life, as I have never used it in mine. But the guards are shaking down cells and we do what we must. Enjoy the shit show.

NO SHIT, THERE I WAS...A BOX OF GRID SQUARES

This is less of a 'war story' and more of a cautionary tale. It's a common trope that young Lieutenants are always lost. I'm sure that it happens often enough, but I feel duty bound to defend their honor in their absence. Lieutenants are new Soldiers, and soldiering can be complex and confusing. So it's fun to send the 'new guy' on fun little treasure hunts like searching for blinker fluid, canopy lights for a night jump, collecting air samples from motor pool tail pipes, and left-handed screwdrivers. For the uninitiated - these things don't exist. You can have a hapless 'new guy' wandering around all day in search of these non-existent items if everyone is in on the joke. Fun for everyone, sometimes even the 'new guy.'

Some things do exist, but not in the context that they are presented on the treasure hunt list. I've seen more than one hapless 'new guy' get scuffed up pretty good for asking the crotchety old Platoon

Sergeant if he could sign out the "AN-PRC-E7 Radio System" for some training event. For those that don't know, the *AN-PRC* prefix is a common nomenclature for older radio systems. The PRC-77 was the standard issue radio for the military through much of the 70s, 80s, and 90s, so it's not too far from the truth. And *AN-PRC* is most commonly pronounced "prick." The term E-7 is the enlisted grade designation for a Sergeant First Class in the army. A Sergeant First Class is the common rank for a Platoon Sergeant. So sending a new guy to ask the crotchety old Platoon Sergeant if he can borrow the "Prick E-7" is certain to elicit a smoke session.

During field training, the young Lieutenant is often consumed with his map. It's a bad mark to get your platoon lost in the woods, and whenever a platoon is 'temporarily disoriented' the Lieutenant is certainly going to catch the blame. So there is an entire sub-culture of jokes, pranks, and tomfoolery surrounding officers, maps, and getting lost. I've seen more than one 'new guy' fall for the old joke of, "Hey, look at the map and tell me if you can see me shaking this tree real hard," when trying to locate themselves. Sixty percent of the time, it works every time. Don't fall for this one. You've been warned.

But the tired trope of officers not knowing how to navigate is not well supported at SFAS. Land navigation training is a significant part of most commissioning processes and while I have seen some officers who simply can't navigate, at SFAS officers are about twice as likely to pass land navigation. That's right, officers are better at land navigation. This is empirically proven and there has never been a class that officers did not do better than their enlisted counterparts.

But don't let facts get in the way of a good joke. I've got your proverbial 'box of grid squares' right here. They're next to my training certificate that tells me I got Selected.

3

CALCULATING DISTANCE

"Everywhere is within walking distance if you have the time." – Steven Wright

How far do I have to walk? That's really at the center of this calculating distance process. When you're plotting distances on a map it's sort of theoretical in nature. It's just ink and paper. It's only a few inches, right? But in reality, we are building to some practical application...how far do I have to walk? With a rucksack on my back. In the rain. At night. With a big nasty blister on my heel. Oof. In this context, calculating distance catapults well beyond the theoretical. Being able to accurately calculate distance on a map and translate that into how far I have to walk is now a matter of life and death. Or at least it certainly feels that way on Day 5 and mile 100.

Let's also remember that we are still learning how to read a map. In the coming chapters we'll discuss actual land navigation, but for now we're just on the map reading stuff. We must learn this stuff first. Basics build champions, and all that. So let's get focused on measuring map distance first and deal with the walking when that time comes. Because our map follows a standardized scale (1:50,000 in this case) we can very accurately measure distance. The bar scales located at the bottom center of the marginal data will be our primary tool, but there are others.

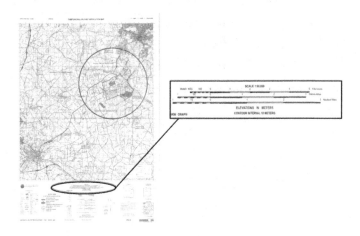

This is where the bar graph lives.

Let's Go to the Bar

Here is the bar scale in detail. There are 4 pieces of data in this graphic, but only 2 (maybe 1.5) are useful to us. Let's eliminate the less useful ones first. Mile and Nautical Mile are first on the chopping block.

Let's be clear...I'm an American. A red-blooded, freedom-loving, mile-using American. But only when I'm in my car or giving directions. Because this is what red-blooded, freedom-loving, Americans do. I also know that this near enrages our Euro-trash cousins who struggle to acknowledge American Exceptionalism, so I keep miles in my bandolier for when I fire my freedom musket across the pond. Most people don't recognize (especially our freedom-starved Euro-trash cousins) that the metric system is the official US standard. That's right, in 1975 the US signed into law the Metric Conversion Act of 1975. It's official. The US is metric. But because we are American and it is our nature to be rebellious, the Imperial system still prevails. The flagpole on the moon that carries the Stars and Stripes is measured in feet, not meters (it may actually be in meters, but you get the metaphor!). But for Land navigation, we use meters.

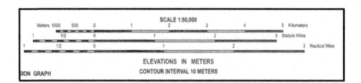

The Bar Graph

So what are miles and even more perplexingly nautical miles, doing on the map? Nautical miles are easy to explain. You may recall our earlier discussion of latitude and longitude and its role in early navigation (and modern navigation for you Navy weirdos) and establishing a grid reference system. You may also recall that lat/long coordinates are expressed in degrees, minutes, and seconds. The equator is a circle (which we have already established has 360 degrees). Each degree is divided into sixty minutes, which are not the same as the minutes on your watch. In nautical navigation, one minute is called a nautical mile. So each degree of latitude is sixty minutes or sixty nautical miles. It's a mathematical calculation based on degrees of latitude around the equator. It's an old holdover, but it remains on the marginal data. A nautical mile is 6,076 feet.

A statute mile, or regular mile is 5,280 feet. It's called a "statute" mile because its length was determined and decreed by an English act of Parliament in 1592, in a *statute*. A statute mile is about 1,600 meters. There are many other miles...an Irish mile (2,048 meters), a Roman mile (1,479 meters), a Danish mil (7,532.5 meters), a Norwegian and Swedish mil (anywhere between 6,000-14,485 meters), an Arab mile (1925 meters), and even a Portuguese milha (also used in Brazil) coming in around 2,087 meters. We should also note that there are no Irish, Roman, Danish, Norwegian, Swedish, Arab, or Portuguese flags planted on the moon. American Exceptionalism. So miles, both statute and nautical, are noted on the map margin bar scale, but we won't be using them for land navigation.

We'll talk contours in a moment but let's focus on meters for a bit. Military Land navigation is a meters market. You already went through several lessons where we described and practiced the MGRS and grid plotting process, an activity that uses meters. The entire grid system is predicated on meters, so it's not a large leap of faith to use it for measuring distance. The bar scale accounts for this and the top bar is the meters bar. You will also note that the bar is divided into six sections. The five sections on the right of the bar scale are delineated in kilometers, or 1,000 meter increments, while the left-most section is further divided into 100 meter increments. The same way that you were able to "slice the onion very thinly" in acquiring an 8-digit grid by extrapolating the final numbers on the coordinate scale, you can similarly extrapolate increments smaller than 100 meters when measuring distance. But I find that 100 meters of fidelity is usually sufficient for navigational purposes.

Measuring Straight-line Distance

As such, measuring distance should be becoming apparent now. You have a scale, the bar scale, that delineates distances right on the map. All that you need to do is come up with a way to transfer the bar scale to the particular position on the map that you wish to measure. You can simply do this with any straight line or straight edged implement, like a piece of paper. For these next few map graphics we'll keep the bar scale, appropriately calibrated, attached to the map. I'm going to walk you through, very deliberately, how you can measure distance on a map. We'll start with a very simple straight-line distance of our

6-way intersection to Lake Bagget that we plotted an azimuth for already.

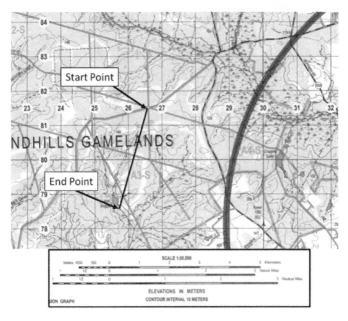

The thing we are going to measure...the distance between our start point and our end point.

The theory is that you can simply "use the bar scale to measure the distance," right? Now even an idiot can understand that this simply isn't practical. The bar scale is printed on the same map, so some convoluted 'origami magic' exercise where you fold the bar scale into position just isn't realistic. But we can use that visual for demonstration purposes as we build your understanding of the technique. So in our instructional fantasy world where we "use the bar scale to measure the distance" it might look something like this.

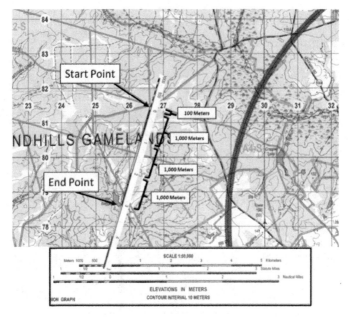

Origami Magic...

You can see that we simply 'clipped' the kilometers bar scale and positioned it along the straight-line between the start point and the end point. There are 3 full kilometers and a little bit left over. This is what the end of the bar scale – the little delineated section – is for. You will recall that the bar scale is subdivided into 100 meter sections on the left end. The entire bar scale is 6,000 meters – 5 x 1,000-meter increments and 1 section divided into 10 x 100 meter increments. You simply slide the scale so that it lines up and count the sections. In this case we have 3,100 meters. That's the distance between the two points.

But we did some 'origami magic' by clipping and pasting that bar scale to determine that distance and that's not practical for us. We need another way to compare the distance that we want measured to the bar scale. We need a facsimile for the 'origami magic' bar scale and fortunately all we really need is a piece of scrap paper with a straight edge. That's it, just a piece of paper with a straight edge long enough to measure between the start point and the end point. You lay that straight edge so that it lines up between the start point and the end point (just like we did with the 'origami magic' bar scale) and then make a tick mark on those spots.

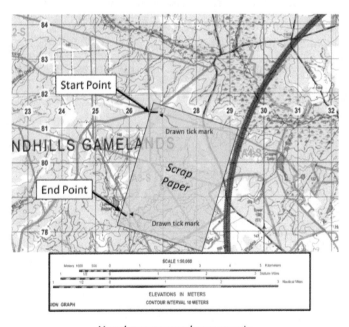

Now the scrap paper does our magic.

Now we have a measuring stick of sorts, and all we have to do is now take our scrap paper (I've made the one in the graphic slightly transparent for demonstration purposes) and line it up with the bar scale on the bottom of the map. We can now use that measurement, compare it to the bar scale, and our distance is calculated. We can see that, unsurprisingly, the distance is 3,100 meters. All you have to do is make your tick marks in the right spots, drag your scrap paper to the bar scale, line the tick marks up, and then count the distance. 3100 meters. Again, you can see how precision matters. If you make your tick marks slightly off the mark or you use a fat marker or crayon, then you might get a slightly different distance. If your tick mark ends up between two measurements you might have to make a "guesstimate" of which measurement is better or split the difference, say 3,150 meters. But the process, the technique, is the same.

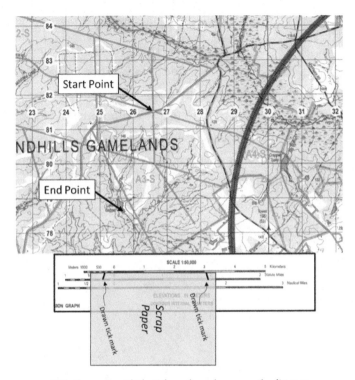

Slide the paper with the tick marks and measure the distance.

Measuring Curved-line Distance

Again, there is nothing advanced about this technique (there really is nothing "advanced" about land navigation in general), this is simply *the* technique. But this is just for straight line distance, which is usually good enough for navigating purposes and you can divide non-linear routes into shorter straight-line routes. But there are often times when we will be required to measure the distance of a non-conforming curved line, say a road for example. How do we measure the distance of a curved line like a road? The same *scrap paper* technique applies, but now we have to shift the scrap paper along the route and pivot the paper every time the straight edge diverts from the measured road. This process is made more precise by placing a tick mark on both the paper and the map to use as a reference index for realigning the paper to continue measuring. Let's see how this might look on a "simple" example before we go to the map.

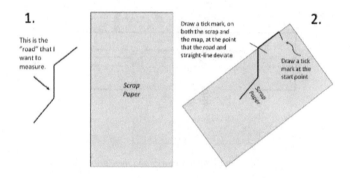

An orderly, predictable, and repeatable method...

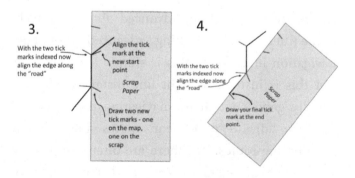

...step by step...

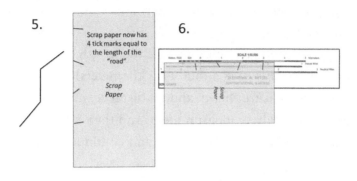

...by step.

We have a simplified "road" and a scrap of paper. We start by aligning the straight edge of the paper along the "road" and making a tick mark at the start point. We make another tick mark where the straight edge of the paper and the "road" no longer align and then make a corresponding tick mark on the map. Now we know where we need to realign, or index, the scrap paper and the map once we rotate the scrap paper to continue measuring. In this case keeping the straight edge of paper aligned with the "road" is easy, because the "road" isn't really curved rather it's a series of connected straight lines. But we did that purposefully for demonstration purposes, and the concept is the same. We then simply repeat this procedure – make a tick mark and keep adjusting the paper along the "road" until you have measured the entire length that you desire. Finally, you now take the scrap paper (with its series of tick marks that measure the full length between the first and last mark) and now lay that down along the bar graph. You can now accurately measure that total distance.

If the total distance exceeds the length of the bar graph (in this case a total of 6 kilometers) then you simply measure the maximum distance, make a tick mark at a whole measurement, and then realign the scrap paper on the bar graph and continue measuring. In our example this process was fairly easy because our "road" wasn't too 'curvy' (only 4 tick marks) and the total distance wasn't long enough to merit repositioning the scrap paper. If you are measuring a very 'curvy' road then you might have 50 or 60 tick marks. Every time the straight edge of the scrap paper and the curvy edge of the road deviate you have to make a tick mark (remember our discussion about precision and line thickness?). In the interest of some brevity and clarity, let me show you a start position and an end position of a curvy road measurement.

Let's take this exact same process and apply it to a road on our map. We'll pick a "clean road" without too much curvature or clutter to make the demonstration a little clearer, but the process is exactly the same. Grab a scrap of paper with a straight edge, make a tick mark on the scrap and the map at the start point, and just follow the road making corresponding tick marks where the two lines deviate until you get to the end. Then you simply take your scrap paper with all the tick marks and measure it against the bar scale. It should look something like this when you're done.

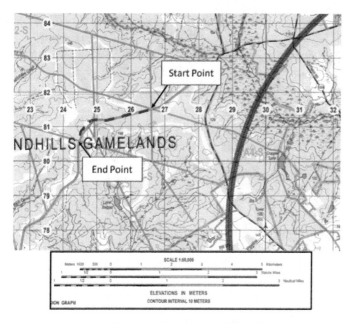

An orderly, predictable, and repeatable method.

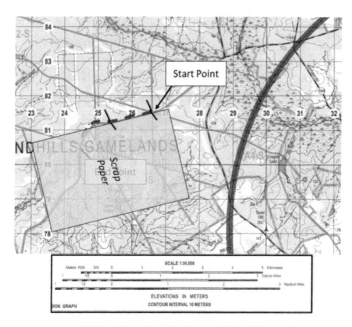

Follow the road and make your tick marks...

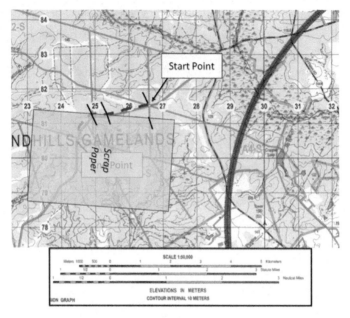

Keep following...keep ticking...

Keep following...keep ticking...until you get to the end point

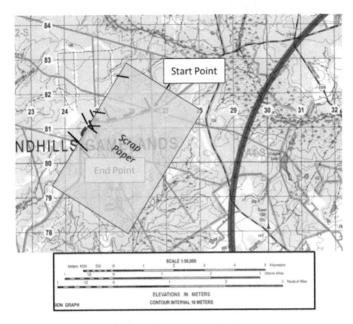

Use the bar scale to measure the distance.

This is how you measure distance on a map. Eventually you will end up walking that distance, but this is how you accurately measure it on a map.

NO SHIT, THERE I WAS...THE GHOST HEADLAMP

There exists an eternal struggle between Cadre and wayward Candidates. Candidates try to push the limits of the rules and Cadre endeavor to enforce standards. There really aren't too many rules, but Candidates have an uncanny ability to push against them in the most creative ways. One of these rules is that Candidates are not authorized to use a headlamp while moving. Every Candidate is authorized a headlamp (two are actually required on the SFAS Packing List. Make sure to read our continuously updated Packing List Manifesto posted on TFVooDoo.com where we give detailed analysis of every single item on the packing list along with our recommendations for the best gear to bring with you to SFAS.). So you are authorized a headlamp, but you are only authorized to use a red lens headlamp and only while stationary. The idea is that you can stop periodically and turn on your headlamp to check your map and such. The intent is to restrict you from walking around and navigating by the light of your headlamp. Some classes are told that they can only use their headlamp while under the cover of their poncho, which is the correct answer for a tactical environment. But the rule is generally no white lights and no headlamps on the move.

I was out one night with Cadre monitoring training, and we had positioned ourselves at the top of a little hill leading into a fairly thick draw. For those that aren't familiar, a draw is a type of ravine that is usually marked by sloping sides and thick vegetation, usually with a swampy water-logged area in the bottom. At Camp Mackall the draws can be quite aggressive. At any rate, we were sitting at the top of this hill looking into the draw and we spotted a white light slowly moving through the thick brush. This is of course against the rules. So we started slowly moving to the wayward light source. Not wanting to jump to conclusions, we wanted to ensure that it wasn't an injured candidate or some other crisis unfolding. Cadre aren't malevolent, but they aren't fools either.

After a few minutes, it was clear that this particular light source wasn't under duress and was simply using his light to extract himself from the tangled brush and marshy footing. We were now stationed at the edge of the thick brush, maybe 100 meters from the candidate, and the Cadre called out, "Candidate, what are you doing!?" The light suddenly stopped moving and the Candidate meekly replied, "...Uh, who? Me?" The Cadre, not satisfied with the response called back, "Get your ass up here and report to me!" The light suddenly switched to red, but it was too late. He was busted. We moved out of the brush and waited for the Candidate to report.

As we moved up the hill a bit, we could see the red headlamp bobbing along, the Candidate making his way out of the draw. We waited. The light bobbed. We waited. The light bobbed some more. The Cadre, now growing inpatient called back out, "Hey Candidate, hurry your ass up!" No response. He called out again, "HEY CANDIDATE!" Nothing. The Cadre, now legitimately pissed, set off into the thick brush to provide "direct supervision and motivation" to this Candidate. I followed, eager to see how this would play out. As we walked up to the bobbing light, we realized the gig was up. The light was hanging in a tree branch, and the Candidate was gone.

This Candidate had quickly realized his precarious position, flipped on the red lens, and hung his headlamp on a little sapling. The light breeze that night kept the light moving convincingly, while he darted back into the draw to divert around us. Amazing initiative, but really poor judgment. I should remind aspiring Candidates that you will have multiple real-time GPS trackers on your person and it's a fool's errand to cheat like this. I don't know exactly what happened to this particular Candidate, but I'm certain that he was sussed out and brought to justice. I won't pretend to not be at least partially impressed by his quick thinking, but his decision making does lack some rigor. Best for all that he didn't make it.

And in the end, he also gave up a pretty nice headlamp.

4

ELEVATION AND CONTOURS

"The whole art of war consists of guessing at what is on the other side of the hill."
– Duke of Wellington

Now that we have covered the distance part, the portion where we have to walk, we should talk about what we are walking on. The elevation. The hills. This is a lesson that is, again, mostly lost on paper. On paper, the elevation is just a slight thing. On paper, the hills are just little faded brown lines on the map. They barely mean anything. They mostly go unnoticed. This is true for navigating at SFAS, as the terrain is mostly flat. I often describe the terrain as "gently rolling." That's accurate. So many aspiring candidates discount elevation. But there are some advanced land navigation courses where the contour lines mean a great deal. This is certainly true in the mountains of

Afghanistan. I've patrolled along the Andean Ridge in Latin America where the contours lines are so close (meaning the terrain is so steep) that you damn near can use the lines as a staircase (yes this is hyperbole...don't get confused). So while contours aren't super impactful at SFAS, they are important elsewhere. And a smart navigator can still use the elevation to assist in his route management whilst navigating. Let's dig into contours.

Contour Lines

Not all maps have contour lines, so many folks are unfamiliar with what they are or what they depict. A standard road map is unlikely to have contour lines on it. Contour lines are the little squiggly lines on the map that depict elevation. They are often used as the universal marketing symbol for "outdoors" and "tactical." We've used them on this book cover, for example. Absent of any other features, like vegetation, roads, or structures, the contour lines might look something like this generic "contour line" graphic.

Contour Lines

You may note that some of the contour lines are drawn a little darker than the others. There's a reason and a name for these lines. The darker lines are called *Index* lines and usually designate large whole number measurements, like every 100 meters. These are complemented by *Intermediate* lines that designate the transitional measurements between *Index* lines and is established by the contour interval. So if the Index lines are 100 to 200 meters, the Intermediate lines would measure between these elevations. If the contour interval was 10 meters, they would represent 110, 120, 130 etc. and there would be nine *Intermediate* lines. If the contour interval was 20 meters, they would represent 120, 140, 160, and 180 and there would be four *Intermediate* lines. Finally, there are Supplementary contour lines, represented by the dashed line, and are used to designate sudden changes in elevation.

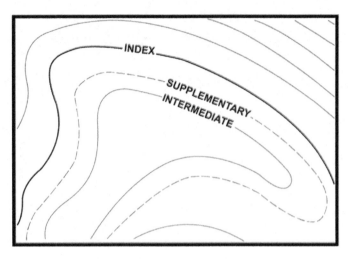

Index, Supplementary, and Intermediate Contour Lines

Rarely will you use the actual elevation designation unless you are using an altimeter, in which case this can be a valuable tool. But at SFAS and most other training environments you won't have this available. So just understand that contour lines represent elevation. But on a military map the contour lines can appear a little less prominent as they almost get lost under all of the other "stuff" on the map. The green vegetation, the water features, the roads, and buildings are what catches your eye on a map. That stuff takes prominence, and the contour lines are just sort of there. And unless the elevation is particularly steep, the contour lines aren't very dense, so they don't jump out as much. This can be punctuated by the contour interval. You will note on this example map that the contour interval is 10 meters. What this means is that when the elevation increases (or decreases depending on your perspective) 10 meters another line is drawn.

You can see from a quick glance at the map that there are entire grid squares where you don't have a single contour line or maybe just one or two. This means that the depicted area is very flat. Look at the 2681 grid square where our 6-way intersection is located, and you can see how there is virtually no elevation to speak of. As I highlighted earlier, most of the Sandhills of North Carolina, where SFAS is located, is gently rolling.

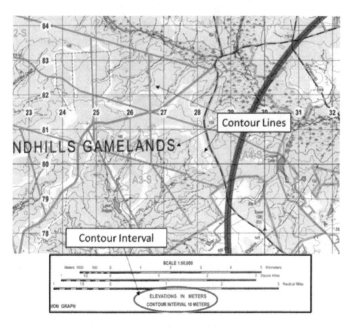

Contour lines on our actual map and the contour interval.

But what those little squiggly lines represent is actual elevation. What those lines represent are little slices into the terrain. Each line is like a cut. The cut pieces of terrain stack up on top of each other and make the terrain feature. In order to see what that looks like, lets isolate a single hill. From overhead, a top view, a hill looks like a series of concentric circles. If the circles are perfectly round and centered then the resulting hill would look like a ball cut in half. If the circles were slightly oblong and off centered then the hill would mirror that shape, as in the graphic below.

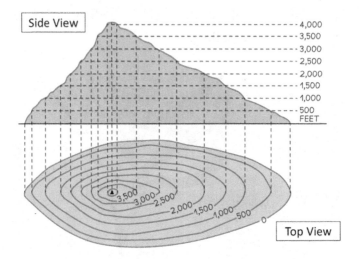

What the contour lines represent.

You can see that on the left side of the hill the contour lines are closer together in overhead/top view. This results in a steeper slope. While on the right side of the hill the contour lines have some more space between them and this results in a more gradual slope. So, lines close together is steep, lines farther apart is less steep. The intensity or severity of the slope is determined by the *contour interval*. On our example map this interval is 10 meters. On the above graphic the interval is 500 feet, which is significant. That's why it's used for the graphic demonstration because the slope is so noticeable. That is a steep hill to climb.

Back to our regular map. We've already discussed the relative lack of elevation depicted in this area. There are few contour lines, and the contour interval is only 10 meters. But there is still *some* elevation, even if it isn't very pronounced. When you are out walking, you can still see (and certainly feel) when you are going down (or up) a little slope. Because the area depicted is so flat, when there is some slope you usually see water gather at the bottom. That's seems to be one of the defining features of this area...the low-lying swampy areas. When terrain is steep, the water sheds quickly and drains rapidly. It usually doesn't present as a swampy area because the water doesn't linger long. In these generally flat areas depicted on our map, the water lingers. You can see this in the 2779 grid square. Look at that grid square and notice both how there is little elevation, but what elevation there is allows that water to linger. The result is a preponderance of swampy area. We'll talk about the major and minor terrain features in a moment, but these low-lying swampy areas, usually accompanied with upwards sloping terrain on two or three sides, are called draws. Where there is water there is usually thicker vegetation. Sloping sides, swampy waterlogged ground, and thick vegetation.

As such, a skilled navigator will note when the terrain is sloping up or sloping down (even slightly) and can use this minor change in elevation to help him navigate. It can serve as an enroute rally point (which we'll cover later in the book) and highlights one of the foundational navigation skills – terrain association. This is especially true at night when your ability to see the terrain is limited, but your ability to *feel* the terrain remains unfettered. Remember this, it will become important. It is mostly lost during this theoretical, paper and ink, comfortably reading about it type of discussion. But it will absolutely be important on the ground, and that's the logical conclusion to this process.

This next graphic reinforces the understanding that the lines represent elevation, and learning to understand what they are depicting can help you to visualize the terrain. This terrain feature is called a saddle. We'll cover all of the major and minor terrain features in the next chapter. Frankly, your experience with these terrain features is likely to be extremely limited. At SFAS you will experience draws and spurs, and draws…and more draws. And another draw. But we'll cover them all in a moment. As you develop your navigation skills on the ground, you'll find that you start to "see" the terrain differently. You will be able to visualize what form the lines are creating and you'll start to place yourself in a point-of-view attitude that can envision how you might move through that terrain. This is a critical element of the skill of terrain association.

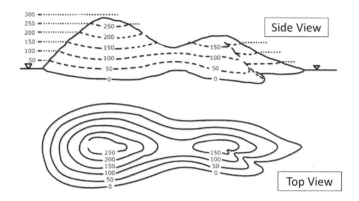

The contour lines help you to 'visualize' the terrain.

Slope

Officially, the Army classifies slope into four categories: gentle, steep, concave, and convex. But that sort of pedantry doesn't really mean much in application. Nobody has ever tested me about what kind of slope I was walking up. I'm still walking on the slope, so don't bother me with classifying it. You'll learn to start recognizing relief, elevation, and the contour lines that depict it on the map. It won't mean much while you're just reading about it and looking at them on the map, but it will mean a great deal when you get on the ground and start to feel its effects. Knowing what the lines represent makes you smart. Understanding how they impact your movement on the actual terrain makes you wise. You'll need both to be a good navigator.

NO SHIT, THERE I WAS...THE DRAW MONSTER NEVER LOSES

Stay out of the draws. Stay out of the draws! STAY OUT OF THE DRAWS! STAY THE FUCK OUT OF THE DRAWS!!! If you remember nothing else, then remember this. The draws are unforgiving. They eat time and devour Candidates. Ancient lore requires that Candidates who enter a draw are required to offer a small equipment sacrifice to appease the Draw Monster. There are more multitools, map cases, hats, gloves, and more than a few weapons sitting in the draws of Pineland. I have personally found multiple Rubber Duckies just sitting there in the brush, an unsecured weapon and non-selected roster number written on the side. I'm convinced that you could fully stock a surplus store with all the gear left to the Draw Monster.

I always emphasize the importance of avoiding the draws as part of your route planning process. It's almost always faster to walk a few extra kilometers and circumnavigate the draws than it is to go through them. Some of the draws are so thick that it might take several hours to successfully penetrate them. It only takes about 15 minutes to navigate a kilometer, so a 3-kilometer avoidance route is just 45 minutes. Contrast this with a 3-hour (or more) deep penetration draw buster and the inevitable physical exhaustion, wet feet, and equipment taxing result and the longer, but faster, circuitous route just makes more sense. But its not *always* possible. Despite my admonitions, some routes have no other option but through, and the results are worth noting.

I once saw a Candidate enter a particularly thick draw in hopes of avoiding the 30-minute detour and 30 minute return detour that a Scuba Road crossing would entail, and emerge FIVE HOURS LATER. But that's not even the best part. He entered the draw tracking a north to south heading. He got discombobulated, went 10 rounds with the Draw Monster, rallied to make his movement, and emerged…except he was traveling south to north. He got, quite literally, turned around whilst battling the demons (both in his head and in the mud) and never got through the swamp. He unwisely entered and got pushed back the way he came, and he was so bedraggled that he stopped checking his compass. So not only was he 5 hours behind schedule, but he also still had to get across the draw and find his point. And the cherry on top is that when I caught up with him, I noticed that he was traveling a little light…he didn't have his weapon. He was so exhausted and embattled that he hadn't even noticed it was gone.

Some guys make good use of their weapons. You have a non-functional 10-pound stick that makes a ready quasi-machete when pressed into service. I've seen plenty of Candidates choke up on the barrel and full-on homerun swing at the vines to clear a path. I've seen desperate Candidates so deeply stuck that they have to stand up, turn around, and forcefully ruck sack flop backwards just to push down the vines and make a few feet of progress. Crawling, literally crawling on hands and knees, is an option if you can find a game trail to follow. But all of these techniques are exhausting. They sap energy, tax your body, and perhaps most importantly take too much time.

A few years ago, there was a rash of guys with chainsaws and forestry marking tape that were cutting trails through some of the draws. There were near highways through some of the likely routes. But I think that's wasted effort. You have no way to forecast what routes you'll get assigned and its pretty easy to just plot better routes. The likelihood of you getting a non-navigable route that you actually cut a trail for, are astronomically low. SCWS has tried restricting access to the training area, but it is located on public lands and the lawfulness of those orders is dubious at best and only effect those that are actually assigned to SWCS. Any private citizen can visit the training area anytime they choose. There is good cause to restrict access, but there is zero cause to create an unenforceable order when you can just provide good training instead. That's always been my focus.

STAY OUT OF THE DRAWS!!!

5

THE TERRAIN FEATURES

"Believe the terrain, not the map"
– Brian Kernighan

The terrain features. Aaaah, the terrain features. This is where the rubber meets the road. Or does it? Everyone likes to cite the five major terrain features, the three minor terrain features, and the two supplemental terrain features as some wisdom of the ancients. As though they are a grizzled Scout out on the Great Plains helping the Cavalry track a rogue band of Indians. And the really pretentious ones will almost always use the classic 'hand demo' to help visualize the features, shown below. I'm only including it here because I've been conditioned, from years of exposure, to showing it to you. It somehow feels incomplete to not include it. Even though I've never really found much utility in it. Maybe it's my learning style or some cognitive deficiency, but for me the learning comes on the ground not in the hand (or the map). But here it is...

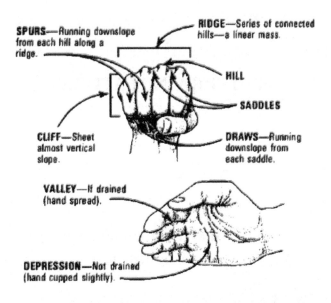

The idea here is that you can hold your hand out in a fist and demonstrate the various terrain features. It's a neat little party trick, but I'm not convinced of its efficacy. It's probably helpful for those moments when you're taking a written land navigation exam and you're asked to recount the features, but it's not super useful in application. We'll keep it in our vernacular so as not to anger the land navigation deities, but we are much more interested in application. I want to know what this stuff looks like in real life, both on a map and on the ground. Let's continue our map study.

All of the terrain features are derived from a complex landmass that we call a ridgeline, a mountain of sorts. As such, if you are in an area far from a mountain range you are unlikely to see all or even most of the terrain features. And the naming convention doesn't make much sense, as we'll discuss below. But because were doing military land navigation and the military likes to be weird with some stuff let's review the weirdness. By the way, the term *ridgeline*, the name of the landmass from which we derive all of the other terrain features, is different from ridge... which happens to be one of the terrain features. Weird, just weird. At any rate, here they are:

The Five Major Terrain Features

1. **Hill.** A hill is an area of high ground from which the ground slopes down in all directions. A hill is shown on a map by contour lines forming concentric circles smallest closed circle in the middle being the hilltop.

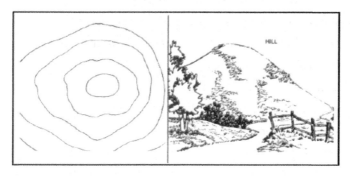

Hill

2. **Saddle.** A saddle is a low point between two areas of higher ground. Most people incorrectly cite a saddle as a low point between two hilltops. This is

incorrect because a saddle could be a dip or break along a level ridge. A better way to think of it is that in a saddle, there is high ground in two opposite directions and lower ground in the other two directions. A saddle is normally represented as an hourglass shape on a map.

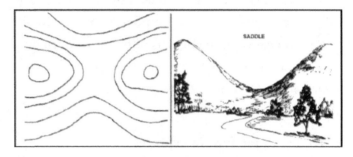

Saddle

3. Valley. A valley is an elongated low-lying basin in the land, usually formed by streams or rivers, usually with high ground on two sides, but sometimes with three sides of high ground particularly when they are just starting. Depending on the size of the valley and where a person is standing (at the head of the valley or in the middle of a miles long valley?), it may not be obvious that there is high ground in the third direction. Generally speaking, a valley on a map looks to have contour lines that are either U-shaped or V-shaped. These contour lines always 'point' the direction of water flow (if water is present) making a good facsimile of an arrow pointing in the direction of flow.

Valley

4. Ridge. Not to be confused with a ridgeline, a ridge is a sloping line of high ground. Makes total sense, especially when the official definition also has the exact words of the thing that it specifically is not...ridge and line. Army are smart! If you cross a ridge at right angles, you will climb steeply to the crest and then descend steeply to the base. Most valleys will be accompanied by some ridges as they are normally what make up the high ground. Ridges can be valuable navigational aids as they are easy to identify on both the map and the ground, so following them creates a natural route. Like a valley, with its characteristic U and V shaped lines, ridges usually have these U and V shaped lines. This makes sense as a ridge is simply an inverse of a valley.

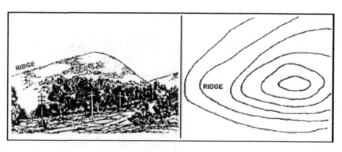

Ridge

5. **Depression.** I won't lie, this one confuses me. We officially classify a depression as a major terrain feature, but I've never actually seen one in the wild. For a guy who has a fairly extensive portfolio of navigational experience across the globe to NOT have seen a MAJOR terrain feature sort of negates the whole *major* element of the title. I don't think I've ever actually seen one on a map outside of a test where you had to actually identify a depression. In other words, they're rare. But apparently it's still a *major* terrain feature. It feels a little bit like the Stop, Drop, and Roll fire safety mantra that I was bombarded with as a kid. It seems like we were constantly drilled on what to do in the event that we were suddenly engulfed in flame. I don't keep super accurate fire watch records, but I'm certain that the number of times that I've had to Stop, Drop, and Roll is definitely less than a dozen. But I'm prepared, so I'll prepare you with a review of the depression.

A depression is a low point in the ground like a sinkhole. I might describe it as simply a hole in the ground. Low ground, surrounded by high ground on all sides (whereas a valley is surrounded on three sides). The scale is usually relatively small, particularly in comparison to a valley. On the map, depressions are represented by contour lines that have tick marks pointing toward the low ground.

Depression

The Three Minor Terrain Features

1. **Spur.** A spur is a short, continuous sloping line of higher ground, normally jutting out from the side of a ridge. A spur is often formed by two rough parallel streams, which cut draws down the side of a ridge. The ground is sloped down in three directions and up in one direction. Contour lines on a map depict spurs with the U or V pointing away from high ground. Its sort of helpful to think of spurs as mini-ridges.

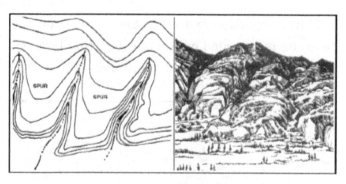

Spur

2. **Draw.** Again, I find myself confused. Minor terrain feature...a draw? Minor? Are you fucking kidding me? Whoever determined these categories must have been a Tanker. I guess sitting in the cupola of a tracked vehicle with a jet turbine engine powering your movement along, might make you believe that a draw is a minor thing. But brother, I'm here to tell you that this classification is the only thing *minor* about a draw. Just go re-read *No Shit, There I Was...The Draw Monster Never Loses*. At SFAS a draw is about the only terrain feature that you're likely to encounter, and there will be absolutely nothing minor about it. A depression, which you never see gets a major. And a draw, which you will not only see but will categorically destroy your soul, only gets a minor. Oof.

Officially, a draw is a less developed stream course than a valley. In a draw, there is essentially no level ground and, therefore, little or no maneuver room within its confines. If you are standing in a draw, the ground slopes upward in three directions and downward in the other direction. They are almost always heavily vegetated, certainly so at SFAS. The contour lines depicting a draw are U-shaped or V shaped, pointing toward high ground. Usually when there are two adjacent spurs there is a draw between them.

Draw

And just so we are clear, here are some draws on our sample map. Learn it, live it, love it.

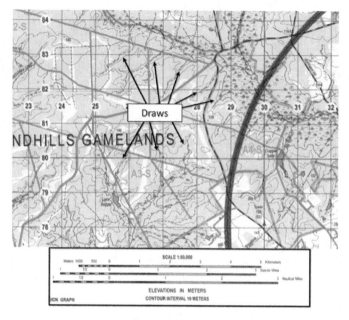

Draws on our map.

3. Cliff. Again, a minor terrain feature that looms large in reality. Although I would note that you very rarely encounter these in the wild, so in that regards you could make a good argument for the minor moniker. But when you do encounter a cliff in the wild it can be a major event, just see *No Shit, There I Was...Minor Terrain Feature.* A cliff is a vertical or near vertical feature marked by an abrupt change of the land. When a slope is so steep that the contour lines converge into one "carrying" contour of contours, this last contour line has tick marks pointing toward low ground. Cliffs are also shown by contour lines very close together and, in some instances, touching each other. You can pretty well envision a cliff.

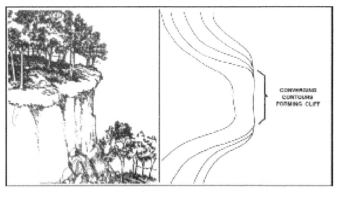

Cliff

Supplementary Terrain Features

Finally, there are 2 Supplementary Terrain Features. Why these aren't called minor and just be done with it, I have no idea. Maybe it's because they are manmade features? But here they are.

1. **Cut.** A cut is a man-made feature resulting from cutting through raised ground, usually to form a level bed for a road or railroad track. Cuts are shown on a map when they are at least 10 feet high, and they are drawn with a contour line along the cut line. This contour line extends the length of the cut and has tick marks that extend from the cut line to

the roadbed, if the map scale permits this level of detail.

2. **Fill.** A fill is a man-made feature resulting from filling a low area, usually to form a level bed for a road or railroad track. Fills are shown on a map when they are at least 10 feet high, and they are drawn with a contour line along the fill line. This contour line extends the length of the filled area and has tick marks that point toward lower ground. If the map scale permits, the length of the fill tick marks are drawn to scale and extend from the base line of the fill symbol.

Cut and Fill

I've navigated in and around plenty of railroads and roadways that are through mountains and rolling terrain, and I know there was ample cut and fill involved in the engineering of these thoroughfares. But I've never actually seen any of this depicted on the map. I've always thought that cut and fill were some romantic throwback to our westward expansion and the wild west. It feels antiquated in a sense, but there you have it. The five major, 3 minor, and 2 supplementary terrain features. They will mean a whole lot more when you start to see them on the ground and you move over, around, and through them. But at the very least, you're all set to take a written land navigation exam. But that's not what land navigation is, is it?

NO SHIT, THERE I WAS...
MINOR TERRAIN FEATURE

It's hard to impress upon most people who have never experienced it, the solitude of a solo long range land navigation exercise. Even SFAS isn't really all that 'solo' as you will regularly see other candidates, roaming cadre, and the training area isn't really all that big. It feels like it's massive, especially when that ruck starts pressing into your neck or when it starts raining. But there are a few advanced training opportunities that truly are solo, long range, and will absolutely tax your navigating skills. I've been to most of them, and they create some of the best *war stories* you could ask for.

I've told some good bear stories before. We recounted *The Magic Knife* in *Ruck Up Or Shut Up* and just remembering that night as I sat by my little fire and listened to the radio still makes me chuckle. And that didn't even involve an actual bear. This one does. In this story our protagonist is Mac. Mac was an experienced operator who was as solid as you could ask for. Cool, calm, and collected. You could trust Mac to spend a day navigating across the mountains, solo, without too much worry. And in this training, you truly were solo. You could navigate the entire day, over 30 plus kilometers, and only see a cadre member at your rendezvous points. Never spot another human and certainly not another Candidate. But you would see plenty of wildlife.

In this episode, as Mac described it, he was being stalked. He was making his away along a ridge (or is it ridgeline?) on his way to a point. As he was walking along, he glanced back over his shoulder at a rock outcropping a few hundred meters back and spotted a dark shape. A bear. Not an unheard-of occurrence, but noteworthy. Cool, calm, and collected – Mac continued on his route. No panic, but a few minutes later he glanced back again. The dark shape was closer. In the Special Operations world, we call this an 'indicator'. An indicator and a warning. Mac trekked on, a little bit faster now.

He went another few hundred meters and chanced another look back at his unwelcome companion. The bear had closed the gap again and it was now clear that he was being stalked. This may sound like a cute little encounter with some Smokey the Bear outcome. But let's be clear. You are alone. The only human for miles. Your weapon is a replica. You are in the middle of nowhere. This is serious shit. Mac, keeping his cool, starts looking for options. He looks to his map and notices a *"minor terrain feature"*...a cliff. Mac later confessed that he wasn't sure how this was going to help him but outrunning a bear in the wild with an 80-pound ruck isn't really an option. So he made a beeline for the cliff.

He found himself running headlong for the cliff with the bear in a trundling pursuit, now within 200 meters. As he got to the *minor terrain feature*, he was faced with a decision, and he wasn't even sure what his options were. So he stood at the top of this cliff, popped his ruck off, grabbed his emergency kit from his ruck, and jumped. I shit you not, he jumped. He evaluated his diminishing options and jumped off the cliff into a copse of pine trees. Miraculously he landed relatively unscathed. But now, he was perched in the top of this massive pine tree, and he was teetering just a dozen feet from the cliff face. And the bear was done trashing his ruck and making his way to the cliff edge. So Mac popped open his emergency kit, grabbed one of his emergency red star cluster munitions, aimed, and fired...at the bear.

I want you to imagine this. A 200-pound operator is clinging to life in the top of a spindly pine tree, swaying to and fro, right off of a cliff. A 300-pound bear is rat-fucking his ruck sack and stalking the wayward climber. The 200-pound dude is now firing emergency flares at the bear. Legs wrapped around the tree, broken branches askew, and a raging bear. And this guy is shooting star clusters at it. AND HE HITS IT! He scored a direct hit and pinged the bear right in the chest and face with a star cluster. Unbelievable.

His lucky shot scares the bear off, but now Mac is still in this tree. So he shimmies down the tree to the base of the cliff, but now he's at the base of the cliff and his ruck is at the top. But it's only a *minor terrain feature*, right? Except Mac has to scale this thing now, fresh off the adrenaline dump of his wildlife encounter. As he later recounted to us, he was simultaneously smoked and exhilarated as he hand-hold and foot-hold made his way up the cliff. All while keeping a watchful eye for the bear, in the unlikely event that he returned. Mac eventually made his way to the top, retrieved his slashed ruck, and finished his point. I wouldn't have believed it if I hadn't personally seen his ruck...and his expended star cluster munition. Do you get a marksmanship award for that?

So a cliff is officially classified as a minor terrain feature, but a cliff is anything but minor. Mac may owe his life to that cliff. Sometimes the bear gets you, and sometimes you get the bear. Mac got the bear.

Authors Note: I know that this story sounds fantastical. It is a war story after all...10% true 10% of the time? The day that I put this story to paper, I had written it down as I had remembered it, and it didn't include the star cluster body shot. I had forgotten all about that part. Later that day I ran into a couple of other Grey Beards that were at that course with me and as we were reminiscing, they reminded me of the pyrotechnics. If you choose the life of a Green Beret, you are bound to see some of the most unbelievable shit. Believe it!

6
INTERSECTION AND RESECTION

"God created war so that Americans would learn geography." – Mark Twain

We're going to dive into some lessons on self-locating that are very important. Not for SFAS, as you won't use them at all at Camp Mackall. But they are important nonetheless as a generalized land navigation and map reading skill. The reason that you won't use them at SFAS is because they are predicated on a few conditions that simply don't exist in the gently rolling hills that make up the Pinelands where you will be training and testing. These techniques require you to be able to see an identifiable object in the distance... a water tower, a church steeple, or a prominent hilltop...and then through a series of measurements and calculations you can determine a location. But this means that you first have to have an object, and secondly that you can see it in the distance which usually requires some elevation. Where you will be training and testing land navigation at SFAS is devoid of these characteristics. There are no identifiable objects, and the lack of real elevation means that you often can't see very far. You are in the proverbial forest with all of its trees.

But we shall learn these techniques nonetheless, because you can't have a book about land navigation and not include them. I should also note, again, that these are techniques that I've never once had to employ in the wild. I've found that in areas where there is sufficient rolling terrain and identifiable distant objects, I never get lost. It's usually fairly easy to stay oriented, even stay oriented to far-off objects. I'm not denigrating the technique, but it sometimes feels like we codified these as the *THE* techniques based on a survey of the German countryside in the WWII era. Those that have been to Bavaria understand that the steeply rolling terrain with hamlets and churches intermittently dotting the view with prominent roads and railways makes for a near picture-perfect Graphic Training Aid that doesn't really exist outside that specific area. Que sera, sera. At any rate the techniques are called intersection and resection, with an additional twist for modified resection.

Intersection

With the intersection technique I am starting with a known location (two actually) and identifying an unknown location in the distance. How it works is that I, in my known location, see an object in the distance. Let's say that I am on a specific known hilltop, and I can see a water tower in the distance. I know where I am, and I can see something in the distance that I want to determine the location of. Simultaneously, another party – say another patrol from your element – is on another specific known hilltop and can also see this "unlocated" water tower in the distance. If I plot my location on the map, and I plot the other patrol location on the map, I now have two points of observation. Then all I have to do is pull out my compass and shoot an azimuth to the water tower. If my other patrol does the same thing, we now have two azimuths from two known start points. When we plot all of this on a map, where these two azimuths intersect, hence the intersection technique, we should have the location of the water tower. It might look something like this.

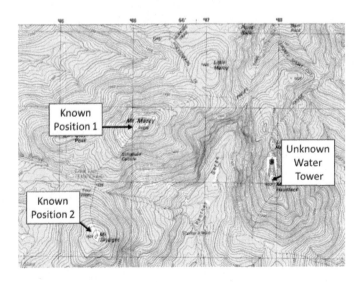

So I'm in Known Position 1 on top of Mount Marcy, my other patrol is in Known Position 2 on top of Mount Skylight. We can both see the water tower off to the east. If we shoot a precise azimuth to that water tower we can use that information (after we convert from magnetic to grid azimuth) to precisely locate that water tower. It would look something like this.

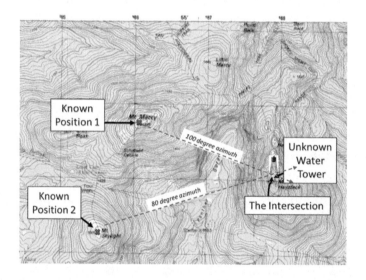

This process can be even more refined with greater precision if you include additional known positions. If, for example, you placed a third observation post on Little Marcy to the north of position 1 and position 2. With this added data point you can triangulate the unknown location for even greater precision. The concept can be expanded to more and more positions ad infinitum, but the process is the same for the intersection technique. Find an unknown location from two (or more)

known positions with their corresponding azimuths.

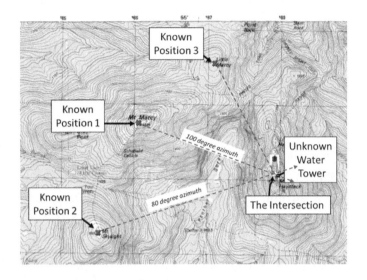

Resection

A resection (and modified resection) is the converse of this process, with some public math for fun. We don't know our location, but we know the location of two (or more) objects that we can see in the distance. Let's just reverse engineer the terrain that we just used for intersection. Let's say that we are somewhere in the valley west of Mount Haystack. We don't know precisely where, but we can look off to the west and see the peaks of both Mount Skylight and Mount Marcy. If we shot an azimuth to Mount Marcy, we determine that it is 290 degrees. We need to plot this on the map now, so we mark the location of the peak (which we can identify both on the map on in the distance) and then plot the "back azimuth."

The "back azimuth" is the azimuth that we shot from our unknown location to our known location (Mt Marcy), but 180 degrees (backwards...back azimuth) difference. So if the azimuth we shot is 290 degrees, then the back azimuth is 110 (290-180=110). If we repeat the process to our second known location, we will get 260 degrees minus the 180 for a back azimuth of 80 (260-180=80). We can now plot each back azimuth from its corresponding known point and where these back azimuths intersect is our unknown location, now known because we just 'found' it via the resection.

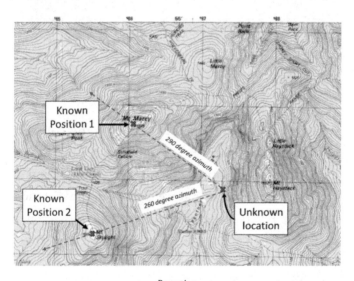

Resection

Modified Resection

There is an additional method called a "modified resection" where we again are trying to locate our unknown position by finding the azimuth (and corresponding back azimuth) from a known location, but this time we are along a linear terrain feature like a road, railway, powerline, or shoreline (river, creek, or large body of water). For this technique we only need one azimuth because the linear terrain feature is in essence our "second azimuth." It might look something like the below graphic. I am somewhere along the creek south of Mount Marcy. I can see the peak to my north, and I shoot an azimuth of 350 degrees. I can now plot a back azimuth (350-180=170) and plot that azimuth from the peak on my map. Where that 170-degree azimuth crosses the creek is my location.

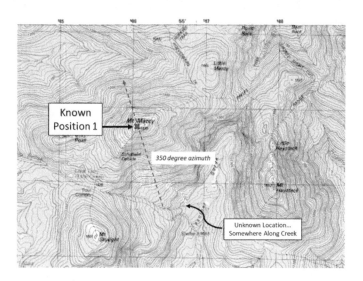

Modified Resection

The one thing that you should note, harkening back to my introductory commentary at the beginning of this chapter, is that the map that we used for the intersection and resection techniques looks very different from the Camp Mackall map. The terrain is extreme. The contour lines are closely stacked and there is significant elevation. As such, this terrain is likely easy to navigate in. The movements themselves are very difficult due to the steep slopes and thick vegetation (the entire map selection is green!), but it's tough to get lost because you know when you're at the top of a hill (or mountain) and you know when you're at the bottom of a gorge. So, with a few azimuths and a little bit of situational awareness, it's easy to stay oriented. This example graphic is an excerpt from a map near Lake Placid, NY. This is mountainous territory, and it is relatively easy to employ these techniques. Contrast this with the Camp Mackall area and you see more elevation change in a single kilometer of vertical ground distance here than you see in 20 or 30 kilometers (or more) of ground distance in Camp Mackall. At selection, there isn't much to see and there's no high ground to see it from. Learn these techniques so that you have the skill, but you won't use the skill at Mackall.

NO SHIT, THERE I WAS…THE CASE FOR MAP CASES

Map cases are a special area of emphasis for me. The topic is well within my "research portfolio" and I approach it from a rigorous academic research perspective. I'm a Geardo with a doctorate, so I weaponize my autism and use it as a superpower. The downside is that I obsess over stupid shit like map cases. The upside is that everyone else gets to benefit from my weaponized autism and enjoy deep analysis of stupid shit like map cases. Map cases probably rank pretty low on the list of stuff that you care about. That's fair. They aren't super critical, until they are. Like if you were attending an assessment and selection course that had you doing land navigation in all sorts of rough terrain and bad weather. *Then* a map case is pretty important.

It was during my attendance at an advanced land navigation training course that this lesson became super clear. This course was attended by relatively senior and experienced Special Operators, but also included a few relatively junior and inexperienced conventional guys as well. One trainee in particular was uniquely junior and inexperienced. He came to this course from the Army Band. That's right, the band. I think he was a flautist. It's tough to get more phallic than that. And some positions in the 'ceremonial units' like the Band and the Chorus have such odd backgrounds. This guy had a Masters degree in Fine Arts and entered the service as a Staff Sergeant. He only had a few years of Soldiering under his belt, and it was Band Soldiering, which is exactly as rigorous as it sounds. At any rate, he showed up to this advanced land navigation course without a map case.

It quickly became apparent to him that this was a mistake. This course is very field intensive, and you essentially live with a map in your hands as you navigate across the mountainsides. We also had the misfortune of bad weather. A map case was an essential piece of kit. So Staff Sergeant Band decided that he would make one. He took a bunch of ziplock bags and cellophane tape and crafted his own. Let me be clear...he used standard issue super thin sandwich bags and regular ol' cellophane tape and built a full-size map case that he would then take into the wilds to protect his paper maps from the elements. He was very proud of his Arts and Crafts creation...he had a Masters degree in Fine Arts, after all! You can see where this is headed.

The following day was a real pisser. We had torrential rainfall, and it was a pretty miserable day out in the woods. But with the *proper gear* and a little determination it was all very doable. The rest of the class had completed the training and was back in the barracks for several hours conducting recovery. We noted that the rain was really picking up and *SSG Band, MFA* had yet to return from training. Eventually, after the rest of us were well-fed, rested, warm, and dry, a vehicle pulled up in front of the barracks. The driver got out and went to the bed of the covered trucked and flipped up the tarp to release the cargo... SSG Band. He dropped down and I can only describe his appearance as disheveled, but that doesn't really do it justice.

He came into the barracks looking as though he had survived a capsizing at sea, with a long trip through the surf zone, and then had to dump his ruck for inspection. He had gear and vegetation hanging off of him, his pants were ripped, and his face was scratched to hell. He was not happy. We saw an opportunity. You have to understand the macabre sense of humor that Special Operators have. Busting balls is an art form and when you see a peer 'limping', you attack. SSG Band was limping, metaphorically and literally. So, being good Special Operators, we poised for attack. He dragged himself into the barracks and we gathered around. He looked up, dropped his ruck and just stood there staring off into space. Someone called out, "Hey brother, how'd that map case work out for ya?" He slowly looked up, dug into his bulging cargo pocket, and withdrew a mangled mess of shredded plastic and sopping paper pulp, entirely unrecognizable as a map and a "map case." He wadded it up and forcefully threw it at our feet in a loud wet splat and screamed, "Fuck that thing and fuck you guys!", and stormed off to his bunk...dragging his gear behind him. Fine art indeed.

So the moral of the story is that stupid shit like sewing kits, and boot lacing techniques, and even fawning over map cases are just little things that don't really matter...until they do. Pay attention to this stuff, because even the little things can be a big deal. My weaponized autism for gear has saved my ass more than once. And I don't even have a Masters in Fine Arts.

A NOTE ABOUT LEARNING

This is the part of the book where we start to "get out over our skis" a little bit. Until now we've had a fairly logical progression of material. We started with a basic map survey and a look at the marginal information. We eased into grids because it was a rational evolution from the map itself. The grid lines were literally right there! Then we went into measuring distance on the map because it made sense seeing as we were already talking about meters during our introduction to the grid squares. We've been slowly advancing through these topics without too many leaps in logic or temporary suspensions of disbelief. It's been easy to follow, thus far. But now we're down the ski jump, we're launching into the air, and were getting a bit over our skis.

Trust me. Just keep reading. It may seem a little confusing at first. That's okay, you're learning. It may not make sense that we covered measuring distance in chapter 3 and now we're getting ready to plot a route where we will be talking about legs and estimated time of arrival and checkpoints. You might be thinking, "How am I supposed to know how far I've walked?!?" The answer is a pace count and we're going to talk about your pace count in later chapters. We're going to talk about it a couple of times. It would have been weird to talk about pace count way back when we talked about measuring distance because pace count is on the ground, and we were still getting comfortable with a map. So trust me. We'll cover pace count just like we'll cover everything else.

We are going to cover everything. We are going to cover some things a couple of times, under a couple of different contexts. Some of it may seem odd at first. Some of it may even seem confusing at first. That's okay, you're learning. And navigation can be a tough topic to understand, and it takes time. That's why this book exists. It's a reference book and you may reference it whenever you want. You can read a chapter, set down the book, and think about the topic that you just read about. You haven't immediately mastered the topic. You have gotten dumber from reading it. No big deal, do some pushups to get the blood flowing and read it again. Make notes. Dog ear and put tabs and highlight stuff. This is a reference book, and some stuff will require deeper study.

But trust me. I haven't forgotten anything, and all will be made clear eventually. Maybe not the first time you read through it. Maybe not until you see a video or two. And maybe not until you attend some in-person training. But you will learn this stuff. Learning is an adaptive and iterative process. Trust the process and keep reading.

7

HOW TO PLAN A ROUTE

"Sometimes the road less traveled is less traveled for a reason." – Jerry Seinfeld

Understand that nothing that I write on these pages is enough to prepare you fully for solo land navigating. You need repetitions. You need to get on the ground with a skilled instructor and fumble around with these concepts. You need practice and you need to get lost. You need to get lost so you can understand how to get un-lost. Learn to never *stay* lost. What does that mean? It means that you need three pieces of information to effectively land navigate. You need to know where you are, where you are going, and how to get there. During virtually every land navigation practical exercise, most certainly at SFAS, you will be given the first two — where you are and where you are going. The start point and the end point. It is up to you to figure out how you are going to get there. That's called route selection.

For most novice land navigators, they take the start point and the end point and they mark them on the map. We already learned how to plot a grid in Chapter 2. Then they draw a line between the two points and connect the dots. Now they lay a protractor over that line, and they calculate a distance and a direction. We learned to calculate distance in Chapter 3, and we learned to calculate an azimuth in Chapter 2. Easy, right? For many, that line becomes the route. If that line says the distance is 3500 meters at a 170-degree azimuth, then by default the route is to walk 3500 meters at 170 degrees. If that line happens to cross over some swampy terrain or a draw, or it forces you to cross multiple roads, or it forces you through thickly wooded areas then so be it. That's the route. If you deviate from that route then you might get lost, right? Look at the straight-line route in the below graphic. Notice that it crosses multiple draws, swampy areas, and obliquely intersects multiple roads. A horrible option.

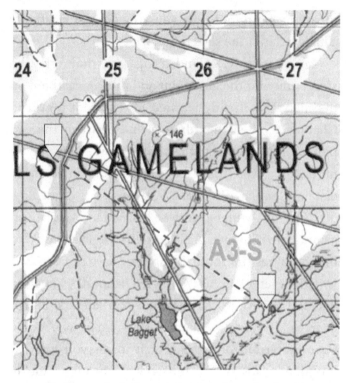

The "connect the dots" straight-line "route"...usually a horrible option.

But that's not land navigation, that's dead reckoning. Just following a distance and direction, absent of accounting for the terrain that actually passes under your feet and around you, is dead reckoning and it unnecessarily restricts you. Land navigation *can* be just dead reckoning, but you have a map that is chock full of valuable information (as we covered in Chapter 1), so why not use it? Why not take the given start point and end point and then do a little analysis to determine how we can use this information to inform us on the best possible way to get there? This is land navigation. In fact, I have successfully navigated multiple land navigation courses without a concrete distance and direction (as described in No Shit, There I Was...Good Job, You're Fired). If you are given the option to pick your route, then fucking pick it! And pick a good one. Here's what to consider.

Route Selection

I like a checklist. It helps me manage cognitive load and if there was an activity that was begging for cognitive load management it would be land navigation. We'll cover this more in the next chapter, but just know that while I pride myself in being a skilled critical and creative thinker, if I can use a checklist, I almost always opt for it. This is true for route selection and here is my 7-point checklist. Let's go through each one, put it on a map, and talk about what they each mean.

Route Selection

1. Start Point and End Point
2. Boundaries and Backstop
3. Obstacles
4. Navigable Terrain
5. Routes
6. Check Points and Attack Points
7. Write it down

Start Point and End Point

Of course we start with the beginning. You are almost always going to be given these two data points...this is where you are and this is where you are going. Once you get to small unit patrolling you might have to do more analysis and figure out these data points on your own, but for land navigation they are almost always given to you. So start by plotting them on the map. Take care to plot them correctly. Take your time, double check that you wrote them down correctly, double check that you plotted them correctly, double check everything. Obviously I've never done this (I allow myself one mistake every fiscal year and this has never been one of them) but I've seen plenty of guys mis-plot an entire grid square over. It's a fairly easy mistake to make, especially under duress or a time crunch. So take your time and plot the start point and the end point.

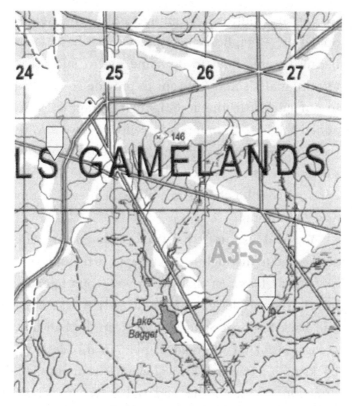

Start Point and End Point - Know where you're at and know where you are going.

Boundaries and Backstop

Once we plot our start point and end point, we have a general idea of where we are going, but we want to give ourselves some left and right limits, literally and metaphorically. We want to set those boundaries literally because eventually were going to have to walk this route and we want to be smart about where were going. Sometimes these boundaries are given to us as training area borders, while other times we can identify them ourselves. In the case of the latter, we should be mindful that we put the boundary there so we can also shift it if we unnecessarily constrained ourselves. Metaphorically we set these boundaries because were about to do some macro and micro terrain analysis and we want to provide some focus to our efforts. In the accompanying graphic I have added the new boundaries in yellow. I have also re-colored the start point and end point (from the previous step on our checklist) to orange. We will continue to color change to orange the previous steps as we go through.

You will also note that I have colored the backstop in red to highlight it. Backstops are features (ideally linear features) that are beyond our intended target, that are designed to do exactly what a baseball backstop is designed to do...keep the ball from going too far off the field of play. In this case, the backstop is a trail that is just beyond (from our likely direction of travel) the end point. When we are making our final approach to our end point and we miss it, we'll know that we missed it when we find ourselves on that trail. This will cue us to turn around and reset our search. Linear terrain features that are oriented perpendicular to our likely direction of travel are ideal as they usually extend enough to catch us if we drift too far in either direction. You might also consider how navigating at night might impact your choices during route selection. Perhaps you need more obvious checkpoints and boundaries. Perhaps you would not brave going so close to that draw at night. We'll cover night navigation in Chapter 8, but consider these things as you develop your understanding of route selection.

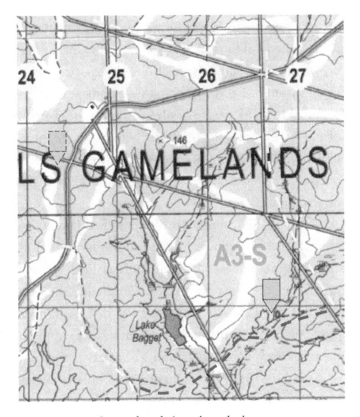

Set your boundaries and your backstop.

Obstacles

The next thing that we plot is obstacles. There is some nuance here as *obstacle* is a relative term. Ostensibly, we are Commandos, so nothing is an obstacle. They told us repeatedly at the Combat Dive Qualification Course that, "Tides, waves, and currents DO NOT effect the Combat Diver!" which is a super cool motto until your ass is being dragged around the bay on a swift tide while tethered to your dive buddies on a bud line. So even though the Army says stupid things like, "Swamps, thickly wooded areas, or deep streams may present no problems to dismounted soldiers..." or "Weather has little effect on dismounted land navigation," (these are actual quotes from FM 3-25.26) I think we can recognize that we need to think carefully about what is and isn't an obstacle. In the map graphic below I have highlighted in red all of the major hydrography (that means water) and some of the more prominent road intersections.

Roads at SFAS are a bit of a paradox. Due to the relative lack of other terrain features (elevation, hilltops, ridges) the roads become more critical. As we'll discuss in the next route selection steps, they are valuable navigational aids. But they are also dangerous. The current rule at SFAS (similar rules apply to most military land navigation exercises) is that you may not be within 50 meters of a road unless you are crossing it at a 90-degree angle. So if you look at the middle of the map graphic and notice the odd and irregular road intersections under the "A" in "GAMELANDS" (that I have heavily marked in red) you see how challenging moving through that area may be if you were following the 50-meter/90-degree rules. Similarly, the 6-way intersection in the top right of the map is certainly very easy to identify and would make an excellent checkpoint and navigational aid, but how are you going to cross that road without violating the 50-meter/90-degree rule? So roads are both good and bad.

The hydrography is less nuanced. You want to avoid it. Where there is water, there is vegetation. Where there is water and vegetation and a little bit of slope there is probably a draw...and that's where Draw Monsters live. I don't know if I've said this yet but STAY THE FUCK OUT OF THE DRAWS! In reality though, there will be some routes that you have no choice but to negotiate a draw. You should do so deliberately and with intent. We'll cover "restricted terrain" procedures in Chapter 9. But for route selection and planning purposes you should avoid water whenever possible. There may be other obstacles that you need to consider. Things like off-limits areas, built up areas, or heavily sloped areas should be avoided, but there aren't many instances of this in most training areas. It is certainly a tactical consideration worth noting, but not so much for our purposes yet.

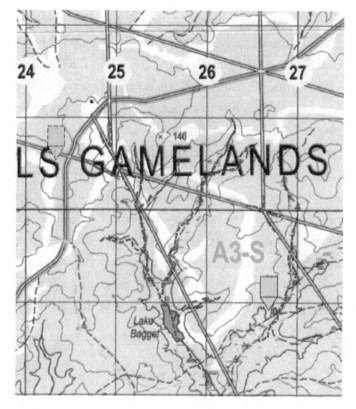

Identify your obstacles.

Navigable Terrain

Now that we've covered stuff to avoid, let's talk about stuff that we can use to our advantage. We'll call this step Navigable Terrain, but you could also call it navigation aids. Essentially, these are things that can help guide us from our start point to our end point. Were looking for things like roads, powerlines, railroad tracks...

> Let's take a quick break from learning for another rant...We often cite railroad tracks as a navigation

aid. They certainly can be prominent, but how frequent are they in reality? Most military installations have them as part of the strategic deployment and redeployment apparatus. You need rail to efficiently move all of that tonnage that an American Army needs to fight wars. But beyond these limited instances you very rarely see them "in the wild." I actually have seen them during a land navigation training event, but they were sandwiched between a large river and a well-traveled 2 lane highway. I was well oriented long before I noted the tracks. I've seen them in the wild once, in a 30+ year history of navigation. So I'm drawn back to my earlier rant about how our land navigation training sometimes seems to be anchored in some nostalgic memory of the WWII German countryside. Hell, our official land navigation manual still teaches Polar Coordinates and Stabilized Turret Alignment navigation while most Soldiers struggle to hold a compass correctly (which we'll talk about in Chapter 9). So maybe railroad tracks shouldn't feature quite so prominently. Okay, rant over.

...we're back to looking for things like roads, powerlines, and railroad tracks as navigation aids. There are others like prominent structures, hilltops, mountain ranges, cultural landmarks, water and radio towers, etc. The problem is that most traditional military land navigation training sites lack these features entirely. This, again, is why roads can play such a large role in staying oriented. Vegetation can also be of fair importance. Pasture and meadows, naturally wooded areas, and managed forests can often be well represented on both the map and correspondingly on the ground. You have to be careful as these features can shift faster than new maps can be updated and produced, but in many cases, these are excellent features. This is true for SFAS. SFAS land navigation is conducted in State Gamelands which are maintained primarily as a managed hunting preserve. As such, there are prominent terrain features called "bowling alleys" that resemble golf course fairways, except they are not nearly as manicured. These bowling alleys effectively create open linear terrain features that transect the entirety of the training area and they are well represented on the map. I have identified them on the map graphic below.

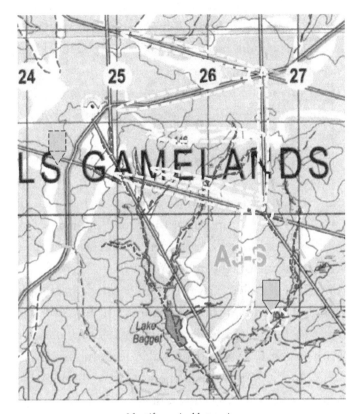

Identify navigable terrain.

Routes

Once you plot the obstacles and the navigable terrain on your map you start to see some patterns emerge. With practice, this revelation will become almost second nature. When properly plotted on a map, these patterns of restricted terrain that we want to avoid and less or unrestricted terrain that has features that are likely to aid our navigation start to reveal likely "avenues of approach." "Avenues of approach" is a common doctrinal term that might not quite fit this application, so I put it in quotes to make that distinction. Remember, nothing about your early land navigation efforts is "tactical," so don't muddy the waters with any of this thinking. Later in your endeavors you will most certainly integrate land navigation with other tactical considerations.

But it's hard enough to learn this stuff well without adding any tactical friction to the equation. So avenues of approach as a non-doctrinal term is good in this description of the broad pathways that get revealed once you plot the obstacles and navigable terrain on the map. You can go back to the nostalgic grizzled Cavalry Scout who can simply sense what the terrain is revealing. In this case, you can see how the middle of our area is very *muddied* with water, draws, and irregular road intersections. The outside perimeter of the area is well populated with roads and "bowling alleys" that create these avenues of approach. It is in these avenues that good routes exist, now we just need to set them.

In this case, from what our macro terrain analysis reveals, we can see that the least restrictive pathways are to bypass that 'muddy middle,' and circle around to the north as a sweeping arc and then dart south once we clear the 'mud'. As a result of this macro analysis, I have sketched a red dashed *preliminary route* to help focus my attention. I could sketch several completely different routes, or branches and sequels from one route. In this case, to demonstrate the technique, I have sketched one singular route. I simply started at my start point, avoided the obstacles, and followed the natural lines of drift and likely navigable terrain and dashed my preliminary route on the map.

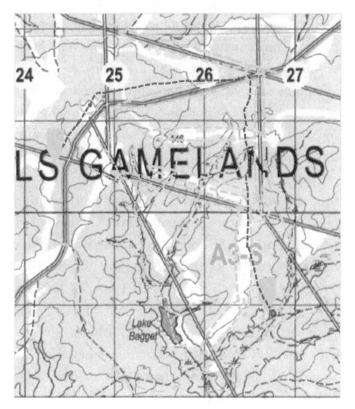

Identify the route

Check Points and Attack Points

At this stage in our route selection checklist, we can start to reap the rewards of the macro analysis that we have done. You can see how we have started in a broad scale with a start and end point and each step incrementally narrows and focuses our attention. We take each element into account, in a deliberate manner, and we end here with a preliminary route that we can now start to micro analyze and develop check points and attack points. Check points are sites along your route that allow you to confirm your location, synchronize your timing, and initiate another phase of movement. The movement between checkpoints (abbreviated at CPs) are called legs. CPs should be aligned with some sort of physical cue. A road intersection, a building, a hilltop, a road crossing, or some physical "thing" that you can identify day or night, so you don't miss it.

If you look at the below graphic and follow the route and the proposed CPs, you can note that they land on sites that meet these criteria. CP1 is sited on the building. We will know when we walk by that building because it's the only building on the map. It lies right next to the road that we can handrail (we'll talk about this term later) from our start point to the building. You simply walk northeast, and you can't miss it. CP2 is the 6-way intersection. This is a prominent feature. It lies directly on the road from CP1 (it actually is the road). You could be braindead and wandering near blind at night, in the rain, and you absolutely could not miss this CP. Walk east from CP1 and don't cross any roads and you'll bump right into it. But, as we discussed in the obstacle section, this irregular intersection also presents a potential danger area. So I want to make clear that **you do not have to occupy a CP to use it as a navigation tool.** You can pass by the building at CP1 without knocking on the door. You can walk towards the 6-way intersection at CP2 without standing in traffic. You can see these things from a distance and still enjoy the positional confirmation. Remember this.

At Checkpoint 2 you make a prominent direction change and head south. You can again handrail the north-south road all the way to CP3. CP3 is a bowling alley. Note this, it is important. Bowling alleys are excellent navigational aids for determining avenues of approach, as discussed earlier. They are linear terrain features that are easy to handrail, but they are also effective when your route transects them because they are so prominent. When you are walking from CP2, handrailing the north-south road, and you happen upon this bowling alley, it will be abundantly obvious. It's a massive open break in the trees, you simply can't miss it. It is a perfect checkpoint, so we made it CP3.

We continue our route south, handrailing the north-south road all the way to CP4. CP4 is another prominent feature. There are actually several features here. There is a significant road intersection. We can layer on top of this a substantial bowling alley. The bowling alley is actually a confluence of two bowling alleys, so it is even more prominent. You would have to be under considerable cognitive distress to miss either one of these features, much less both layered on top of each other. So making this a checkpoint is perfect. That is checkpoint 4, CP4. Our final checkpoint is actually an attack point.

Attack points are unique in that they represent a final positional check before we make our approach to the end point. As such, there are some additional considerations that we take into account. We want attack points to be within 1,000 meters of our end point, even closer if tenable. We want them to be closer because we are likely to start dead reckoning from here, and dead reckoning takes more time. You may have noted during our quick review of our route thus far that we have only been referencing general cardinal directions, not actual precise azimuths. This is deliberate, and good route planning enables you to do this. One of the the factors that I took into consideration when choosing this route is that it had all the characteristics to enable this. One of the drawbacks to dead reckoning is that it is time consuming. I have to be very precise when plotting my azimuth on the map. I have to ensure my protractor is perfectly aligned. I have to make certain that the lines I draw are where they are supposed to be and that they aren't too thick. I have to make certain that I do the math correctly when I calculate the G-M angle declination. Once I'm out walking I have to be careful aligning my compass to the determined precise azimuth. I have to keep it level, I have to be very deliberate in choosing my positional landmarks, and I have to be careful not to deviate from my particular walking route. All of this takes time, so I want to limit the amount of my route that requires me to employ this method. My movement from my attack point is dead reckoning, so I want this dead reckoning final leg to be short. So my attack point needs to be near my end point. The closer the better.

In this case I am using the little pond that is about 200 meters from my end point. But not just the pond, because that in-and-of itself is large enough to induce some error. I am using the precise point where the pond sits on the trail. You may remember that trail from Step 2...establishing boundaries and backstops. That trail is my backstop. So it's a prominent feature that prevents me from going too far past my point, it's easily identifiable on the map and on the ground, it intersects with the pond creating an easily identifiable and precise location from which I can begin my dead reckoning to the end point. All of these little analytical factors and steps are adding up to set ideal conditions for success. It's a complex little dance of understanding the technical procedures, understanding how these procedures translate to actions on the ground, and the wisdom to apply them appropriately. This is why map reading isn't land navigation, and learning to land navigate takes time. It takes repetitions. It means you have to get lost so that you can get un-lost. But this route selection checklist is the perfect procedure to learn the process. As such, CP5 is my attack point.

You should try to select routes that allow you to place CPs every 700-1000 meters. There is no hard and fast to rule to this, but I find that this is the sweet spot. If you are a confident and experienced navigator and conditions are ideal, you could certainly stretch this. If you were navigating through restricted terrain at night, then you might want to have them closer. Remember, the start point and end point are given to you, but you are in charge of selecting the route. So you can elect to have as many or as few CPs as you like, arranged in any manner you choose. Attack points are best a little closer, as discussed. But ultimately, the placement of CPs and attack points are your preference.

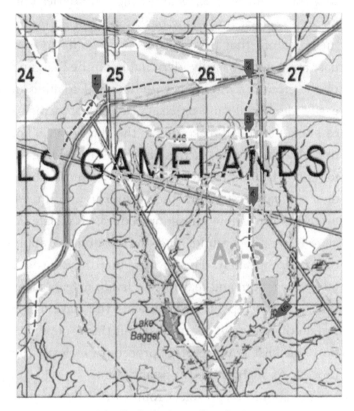

Identify checkpoints and attack points.

Write It Down

The final step in the route selection process is to write it all down. You will recall that the whole reason we have a checklist is for a cognitive load management tool. It would be counterproductive to force ourselves to now increase our cognitive load by memorizing all of the data that we just developed. The grid coordinates, the route, the checkpoints, the attack points, the distances, and the directions. All of the minutiae. That's a lot to remember. So, you should write it down. This way you don't have to risk misremembering it. You can write it down and just reference it along the way. I call this written product the *Route Planning Guide*.

The next graphic is an example of a Route Planning Guide that I use. I teach this at my Muster events, and I've used it for years with great success. You don't have to use this format, but you should most definitely write your data down. It includes the Start Point Location (SP LOC) which is the grid coordinate of my start point. EP LOC is the end point location. I like to have these written down at the top so that in the event that I need to confirm, replot, or otherwise reference them, I have them handy. I next record my Start Point Estimated Time of Departure (SP ETD). This is the time that I start walking. I want this data because, as we will discuss in the next section, it will help me stay oriented. Next, I record the estimated time of arrival (ETA). With these two *time* data points bookending my movement calculations I have some parameters for my movement. The next set of data is the leg data for the various portions of the route from checkpoint to checkpoint. These are called legs. I want to record the distance (in meters), the direction (either a precise azimuth or cardinal direction), and the estimated time of arrival. You may note that this is the third reference to time thus far. We'll talk about this during our discussion of pace. But if I take each of these 'legs' ETA and add them up I can then, based on my ETD, calculate my total/final ETA.

Route 1			
SP LOC:		EP LOC:	
SP ETD:		ETA:	
	Dist	Dir	ETA
Leg 1			
Leg 2			
Leg 3			
Leg 4			
Leg 5			
Leg 6			

Route Planning Guide

The point is that you write this stuff down. I write it down in this Route Planning Guide, but you write it down so that you don't have to remember it. You can choose any format you want, but the end state is that the information is written and thus no longer unnecessarily taxing your cognitive load. I've had students modify this chart and add a column to the Leg section that includes a description of the CP. Just a note that says "intersection," "building," or "bowling alley" is enough to cue your memory to remind you what you're looking for. Whatever works for you works, but remember that you aren't allowed to bring any "land navigation material" with you to SFAS, so a complicated pre-printed route planning chart isn't tenable. It has to be something that you can replicate quickly and easily on a blank piece of paper.

Route 1	Pond Point	
SP LOC: 17S PU 2431 8046	**EP LOC:** 17S PU 2662 7891	
SP ETD: 1230	**ETA:** 3600 M 62 min	
	Dist	Dir	ETA
---	---	---	---
Leg 1	800 M	NNE	10 min
Leg 2	1200 M	ENE	15 min
Leg 3	500 M	S	8 min
Leg 4	500 M	S	8 min
	1000 M	SSE	13 min
	500 M	SSE	8 min

A sample Route Planning Guide for the route we just demonstrated.

Other Route Planning Factors

Your task then becomes to pick the best route possible. Plan the route so that it takes you around the swamp. Select a path that avoids the draws. Instead of constantly crossing the roads, use them as navigational aids. You are in charge of yourself, so take charge. If you get to pick the routes, then why would you purposefully pick a route that forced you to get wet? Maybe the route you plan that skirts around the danger areas is 1000 meters longer. But on the longer route you don't get wet. Maybe the extra 1000 meters takes an extra 15 minutes to traverse, but you didn't walk through a draw and risk getting stuck in that draw for hours. Maybe the route was less direct, but your new route afforded you more easily identifiable checkpoints. These checkpoints allowed you to stay oriented and you didn't have to guess as much. Did you lose time, or did you actually gain some time? Subtraction by addition.

Handrails

Navigating effectively is knowing where you are (start point), knowing where you are going (end point), and knowing how you are getting there (route selection). You are given the first two and you are in charge of the third. You can plan that route effectively by plotting good checkpoints. Your challenge then becomes staying oriented along the 700 to 1000 meters between checkpoints. Do you recall the roads, creeks, and bowling alleys that helped us identify good checkpoints where they crossed paths? Because they are linear, they can serve as boundaries for our movements as well. If you plan your route correctly you can plot that route so that it has one of these linear terrain features on either side of your movement corridor. You are not allowed to be on the roads, but they are going in the direction that they are going, and you can use that to your advantage. If you are moving to the south and a road along your route is oriented north-south, then you can be in the woodline and look at the road and 'follow it," without actually being on the road itself. That is called "handrailing." Much like you would use a traditional handrail to guide you down a flight of steps, you can use the road to handrail along your route. You are not on the road, but you can still use it for navigation. Likewise, your end point may have a linear terrain feature 'behind' it, that is the linear feature intersects your route. In this case, that linear feature becomes a 'back stop' to your movement, that's why we call it a backstop.

Getting *Un-Lost*

Remember I said that you will get lost? It is inevitable. Everyone, even experienced navigators, get disoriented from time to time. If you chose your route correctly and your checkpoints wisely then the farthest that you have to go to get *un-lost* is however far you are from your last checkpoint, hopefully 700 to 1000 meters. When you realize that you are lost and you can't get oriented, then just turn around and walk back to your last checkpoint. You picked them so that they were easily recognizable, they were prominent on both the ground and on the map. So use them! Don't wander aimlessly in the false belief that if you just "navigate harder" your location will magically reveal itself. Be disciplined, turn around, go back to your last check point, and start that segment of your route again. If you "navigate harder" you could be doing so for hours, in the wrong direction. If you are disciplined enough to simply return to your last checkpoint it might take 20 or 30 minutes. Subtraction by addition.

Good navigation then isn't just start point, end point, and distance and direction between the two. It becomes an exercise in checkpoints, handrails, and backstops. Staying oriented along that route. Managing your time effectively. Following all the rules. At Selection, you must do all that with a ruck on your back. Day after day. Night after night. Mile after mile after mile. So plan good routes so you can minimize the suck.

RV Procedures

When was the last time that you left your house without your wallet, keys, or phone? It probably doesn't happen very often, does it? Is this because you have a checklist posted by the door that reminds you? Probably not, but you somehow manage to do this without even thinking about it. How does this happen? I never leave my stuff behind, but I never lose gear at all. I have this weird, perverse sense of pride that I have that I've never, ever lost a piece of gear. Except once. My beloved SOG Paratool. I still remember where it happened and how. It damaged me. It was 1998 at (then) Fort Benning, Georgia. We were laying in a flank security position during an ambush patrol on Red Diamond Road. I had my beloved SOG Paratool dummy-corded to my belt. My ranger Buddy was taking advantage of the lack of supervision and digging into an MRE. He didn't have a knife and he asked to borrow mine. So, I untied the dummy cord and passed it to him. A minute or two later the ambush was initiated ahead of schedule, and we were caught a little unaware. We hustled back to the Objective Rally Point and my Ranger Buddy informed me that he had accidentally left the Paratool up on the road. We obviously couldn't go back for it. Gone forever. I'm entirely convinced that if you took me back to Red Diamond Road, I could find that tree and I could fill that hole in my heart.

But that's not you. You are going to learn what I call RV Procedures, or Rendezvous Procedures. Back to your wallet and keys. The reason that you never lose your wallet and keys is because you have this little subconscious checklist that you do when you leave the house. A little 'pat down' procedure. You just, subconsciously, pat your pockets to make sure that you have all of your stuff. There's no formal checklist, it's just something that you do. Even if you don't physically touch these items, you almost sense when they aren't in their proper place. You step out the door and just instinctually know if something is out of sorts. This little sixth sense is so ingrained in you that it can actually be used against you. Pickpockets will use this to reconnoiter their marks. Take notice the next time you walk out of a bar or restaurant, and you will note that you will often tap your wallet, just to make sure that you didn't leave it or lose it. You are literally telling the pickpocket exactly where your wallet is. It's totally subconscious. Pay attention next time you're out, it's uncanny.

So you have developed this little subconscious practice of patting yourself down to ensure that you don't lose your important stuff. In other words, *you have made routine a vital function so that you don't lose critical items.* Do you see where this is going? I have seen more candidates than I care to recall lose critical items while out land navigating. Everything from scorecards, to maps, to compasses and even weapons and rucks. Yes, rucks. I have seen cognitively overloaded candidates actually start walking off without the ruck on their back. All because *they have not made routine a vital function so that they don't lose critical items.* What we are going to do is develop our own *wallet, keys, and phone* pat down procedure for land navigation purposes. You might call it head, shoulders, knees, and toes. Whatever mnemonic that you use to cue yourself, the end result is that you put your hands, physically touch, each vital piece of kit.

Start at the top – head – and put your hands on your eye protection, your map case (which should be dummy corded and handing around your neck), your weapon, your ruck, everything. Put your hands on it and make sure that you have it. Make a routine. Practice it. Often. Often and always. Just make it a thing that you do so that you don't have to think about it. Hands on your vital gear to ensure it never gets lost. *You must make routine a vital function so that you don't lose critical items.* RV Procedures. Make checking your watch and noting the time one of the steps. If you make this routine that you do, well...routine...then it's just something that you do. You don't have to increase your cognitive load. In fact, a well-practiced RV Procedure will decrease your cognitive load. You won't have to worry about whether or not you have your *this* or you misplaced your *that*. Your RV Procedures will ensure that you always do. Zero cognitive load and your confidence level now goes up. Don't underestimate the power of this routine.

Now put this all together with route planning, getting unlost, checkpoints, and all of the elements that we just discussed. Imagine if you did RV procedures at every CP. What does this do for you? At every CP, which you have deliberately planned every 700-1000 meters, you have a definitive fix on your precise location. You literally planned these locations because they afforded you this advantage. So you know exactly where you are. If you do RV Procedures at every checkpoint then you now know not only exactly where you are, you also know that you have all of your gear. Should a critical Cadre member suddenly drop from the trees and initiate a surprise inspection, you would have nothing at all to worry about. You just passed a checkpoint, so you know where you are. You just did RV procedures, so you know you have all of your gear. You don't have to expend any valuable cognitive capacity on these mission critical conditions. If you get lost all you have to do is go back (no more than) 700-1000 meters to get unlost. If you suddenly discover that you are missing a piece of gear all you have to do is retrace your steps (no more than) 700-1000 meters and you'll find your gear. You are in charge. You decide the route. You decide the CPs. You decide the boundaries and backstop. You decide how fast you need to move, and you decide when you can stop and rest. You decide. You are in charge. This is your mindset.

Land navigation isn't something that you **have** to do. Land navigation is something that you **get** to do. You get the chance to showcase your skills. You get to prove that you can be in charge of yourself. That is powerful. At SFAS, it's required.

NO SHIT, THERE I WAS...
RENDEZVOUS

Many of you readers are likely thinking or planning or already on your way to some military assessment and selection. If you've made good choices then that is likely Special Forces Assessment and Selection, but there are others. I've attended or audited many of them over the years and one constant is the professional tension between students and instructors, or Candidates and Cadre. There is an unspoken, and sometimes spoken and explicit, protocol between the two that Cadre are to remain aloof, disconnected, and dispassionate. I call it being "passionately dispassionate." I care deeply about the outcome of the events, but I can't show you that I care...at all.

Because of this detachment, oftentimes candidate and Cadre interactions can take on a vaudevillian quality. Candidates will get themselves into all sorts of awkward situations and Cadre have to interact, manage, and manipulate the environment as though everything is normal. Sometimes the awkwardness is accidental, and sometimes it's manufactured. Accidental awkwardness is Cadre catching a Candidate crossing Scuba Road in nothing but socks and boots. That's right, butt ass naked, in the middle of a flowing swampy creek bed, holding his ruck over his head. This Candidate wanted to keep his gear dry, and as we saw him slowly emerge from the chest deep water, we thought that he would certainly reveal *at least* some shorts. But as his steps brought him up from the brackish concealment, we discovered that all he had on was socks and boots. And a smile.

Other odd interactions are manufactured. Cadre are playing a role, and it can become a little game to see if Candidates can get them to break character. I had a guy in my class who embraced this stoic-breaker technique. Candidates were instructed to be very deliberate during check-in procedures at the various land navigation points. Approach the Cadre, sound off with your roster number, process your instructions, fill your canteens, and continue to train. Step 1, step 2, step 3, step 4...get in and get out. Efficient. Deliberate. Well-organized.

So this guy decided that he was going to follow the instructions to an absolute 'T.' He approached the Cadre sitting in his vehicle and checked in as expected. He received his instructions and moved to the water cans in front of the vehicle to plot his next point. As he did so, he dropped to a knee and pulled his map out to plan. Then, without breaking eye contact, he pulled out a canteen straw. A canteen straw is a long tube that you use to modify a 2-quart canteen into a hydration bladder so you can drink on the go without actually pulling your canteens out. He stares down the Cadre, unscrews the lid to one of the water cans, and seductively slips his canteen straw directly into the can. He starts aggressively hydrating, all the while never breaking eye contact.

It's a Mexican Standoff...I shall suck this water can dry or you will acknowledge my tomfoolery. When we debriefed him later, he told us that the Cadre never even blinked. He drank damn near three gallons before he had to withdraw. Candidate – 0, Cadre – 1. He slowly slid his straw out of the water can, coiled it neatly in his hand, and slid it into his cargo pocket. Cadre never took his eyes off of him. We can only assume that deep down this Cadre appreciated the engagement. He got Selected and I'm certain that war story made its way around the Cadre..."No Shit, There I Was...this goofball actually pulled out a damn canteen straw!"

If you're not at least having some fun, then why are you doing it?

8

NIGHT NAVIGATION

"Sometimes you have to go through darkness to get to the light."

What follows is the entire text about night navigation from the official US Army Field Manual 3-25.26 MAP READING AND LAND NAVIGATION. The entire text.

∼

1-7. NIGHT navigation

Darkness presents its own characteristics for land navigation because of limited or no visibility. However, the techniques and principles are the same as that used for day navigation. The success in nighttime land navigation depends on rehearsals during the planning phase before the movement, such as detailed analysis of the map to determine the type of terrain in which the navigation is going to take place and the predetermination of azimuths and distances. Night vision devices (Appendix H) can greatly enhance night navigation.

a. The basic technique used for nighttime land navigation is dead reckoning with several compasses.

NOTE: See Appendix F for information on orienteering. recommended. The point man is in front of the navigator but just a few steps away for easy control of the azimuth. Smaller steps are taken during night navigation, so remember, the pace count is different. It is recommended that a pace count obtained by using a predetermined 100-meter pace course be used at night.

b. navigation using the stars is recommended in some areas; however, a thorough knowledge of constellations and location of stars is needed (paragraph 9-5c). The four cardinal directions can also be obtained at night by using the same technique described for the shadow-tip method. Just use the moon instead of the sun. In this case, the moon has to be bright enough to cast a shadow.

THAT'S IT. 230 WORDS. The Army has determined that 230 words are all that is required to document the entirety of its official night navigation guidance. Maybe the fellas that write this doctrine just aren't night owls. Maybe the conceit is that our military just won't move at night. They also note that celestial navigation should be considered. Virtually useless. And of course, the shadow tip method for determining the cardinal directions. More survivalist secret agent man bullshit. The Army never fails to disappoint, and I've seen this laissez-faire attitude wreak havoc in the field. This is no hyperbole – we graduate US Army Infantryman from their MOS producing pipeline without the requirement to navigate on their own at night. That's right, the powers-that-be have decided that 230 words and no outcomes-based assessment of this critical Infantryman's skill is sufficient to field the operational force. No wonder land navigation accounts for almost half of SFAS failures. Despite what the Army thinks, there is some critical stuff that we should cover for night navigation, so let's get started.

Trust Your Sensors

This section is called Trust Your Sensors and what follows is a treatise on all of the various sensors or facsimile of sensors that you should consider. Keeping with our adaptive and iterative approach to learning we will cover this stuff a couple of times in a couple of different contexts. So don't get too distracted when we cover a little compass here and a little compass later. We'll cover the whole compass in the aggregate.

Eyes

The FM does note that the techniques for night navigation are similar to the techniques during the day, just with *limited visibility*. So, let's talk about that visibility. Let's talk about our eyes as a component of the overall systems available to us. These systems, or sensors, collect information that we can use. So it makes sense that your eyes are simply another sensor. In this case, your eyes are the sensor that processes your vision, or how you process light. At night, with limited light for your sensor to collect, you end up with limited visibility. So we can mitigate this by maximizing whatever amount of light these sensors can process.

Your eyes, as a sensor, collect light and process it. This is called photo-reception and is the process that describes how photoreceptors like rods and cones absorb light waves that enter the eye and convert them into electrical signals which are then sent to the brain for visual processing. Both rods and cones are involved in night vision and this phenomenon is known as "dark adaptation." It typically takes between 20 and 30 minutes to reach its maximum, depending on the intensity of light exposure. You can manipulate this time by protecting your eyes from intense light throughout the day and especially in the hours prior to night navigation. In other words, don't stare into the sun and don't look at bright lights. Do you know why pirates wear eye patches? It's not because they all lost an eye in maritime combat. They wear an eye patch to keep that eye sensitive to low light. This way, if they need to go below decks during a hostile ship boarding, they can flip the eye patch up and enjoy better vision to fight the ship's crew in the dark. Flip up the patch and it's like flipping down a set of NVGs.

There is a good deal of science in understanding that the cones are evolved for day vision and can respond to changes in brightness even in extremely high levels of illumination. Cones are unable to respond to light reliably in dim illumination, however. Rods act as light detectors even in extremely low levels of illumination but are less effective in bright light. Both cones and rods are used in dark adaptation, slowly increasing their sensitivity to light in a dim environment. You don't need much more science than to simply understand that you should protect your eyes from too much light exposure.

There are a few techniques that you can use the manipulate this adaptation in the moment as well. For example, you can sometimes artificially increase this sensitivity by squeezing your eyes shut very tightly for about 10 seconds. When you open them again, you're likely to experience a brief moment of enhanced night vision, or at least enhanced contrast. Try it next time your eyes have reached a good level of dark adaptation. Squeeze your eyes shut for 10 seconds and then open them and observe the effect. Test how this response works at various ranges and levels of light intensity. And, because the way that the rods (the structure responsible for low light vision) are generally arrayed on the periphery of the cornea, your night vision is more refined in the periphery of your vision, as opposed to the center. So if you stare at something in low light you might not get the clarity that you would get if you looked at the item in your periphery. What this means in practice for night navigation is that you should keep your eyes moving. If you can continuously scan across your front, you increase the chances that whatever you are looking for will enter your periphery. Don't stare, scan.

Pace

Because your eyes have less sensitivity at night, this impacts your ability to estimate distance. You can't see as much, what you can see is obscured, and your ability to judge how far away an object sits is impacted. So don't trust your eyes...at least don't trust your eyes more than you trust your pace count. Think of your pace count as another sensor and understand that you are not better at estimating distance than your pace is at recording distance. Your pace count is a sensor. Trust your sensor. But also recognize that your pace count at night is likely to be different than your standard day count. For most people, their night count is higher than their day count. They take shorter steps, perhaps representing greater apprehension and caution, while walking at night. But some people (few, but some) take bigger steps at night, so their night pace count is lower. It just depends on the individual and this is one of the reasons why land navigation is just one of those things that you need to get out and practice. Your ability to estimate distance is not as good as your ability to measure it with an accurate pace count. Trust your sensor.

Compass

In the same vein that you can't estimate distance better than your pace count can measure it, you can't estimate direction better than your compass can measure it. Trust your compass. This is especially true given that there are essentially two ways to navigate – dead reckoning and terrain association. And because you can't associate with terrain that you can't see you will be much more reliant on dead reckoning. So you must learn to trust your compass. Which means that you need to learn about how your compass works. So let's get into this. When we are talking about a compass we mean the military lensatic compass. There are many compasses, but this one is mine!

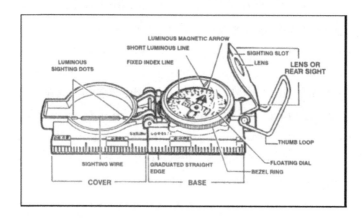

The military lensatic compass is a workhorse. It's heavy and bulky. It's expensive. It can be finnicky. But it is an absolute workhorse. You need to be careful when you buy one. Don't be fooled by a cheap knockoff. The official supplier, Cammenga, makes two versions that look virtually identical. You want the version that has tritium inserts. This is critical. Make certain to check the markings (pictured below) and get the correct version. It is usually twice the cost of the non-tritium version, but there is no comparison. The cheaper version will advertise as 'phosphorescent,' but that just means glow-in-the-dark and requires and artificial light source to "charge" the luminescence. You need the good version and look for one no older than 10 years. The tritium half-life will leave you wishing for a maximum illumination. The date of manufacture is stamped inside the cover and while the manufacturer insists that the reduction in illumination is imperceptible, look for the newest one you can find.

The rotating bezel has a luminous tick mark (labeled on the above diagram as "short luminous line") and is your best friend. Once you properly orient your compass and the determine the correct azimuth, you should align the short luminous line with "luminous magnetic arrow" (which is your north seeking arrow) by rotating the bezel. Now you just need to trust your compass, keep the two luminous tick marks aligned, and keep walking. You are simply not able to estimate direction better than your compass, so trust your sensors.

Drift

Drift while walking is virtually guaranteed. The direction and severity of drift is the real variable. It is usually more pronounced at night due to the reduced visual cues. Most people don't have perfectly equilateral leg lengths and this difference can cause you to drift when you walk. Some people have old injuries, you will get blisters, and the chances of twisting an ankle or tweaking a knee are high while out navigating, so drift is something that you need to account for. Its best to measure this drift on a calibrated course. Generally speaking, you drift towards your dominant hand side. I am right-handed, so I drift to the right. This isn't always the case, but it's a good rule of thumb.

We run a drift course as part of our Red Light Night Series Musters and a drift course is now a semi-regular part of the SFAS enhanced land navigation instruction. It is not uncommon to measure a drift of over 100 meters on a thousand-meter course. Most students have no idea that they drifted and once they learn this and the accompanying mitigation techniques, it can be a game changer. Imagine how far off target you might be during a 5,000-meter leg with this sort of drift. You would be half a grid square off and now you're stuck wandering lost. Learn your drift, then learn how to mitigate it. You need in-person instruction for this.

Dehydration

There is lots of well-researched literature that documents the impact that dehydration can have on your eye health long term and for how it can impact your vision short term. Increased eye fatigue, blurred vision, and more 'floaters' are all common symptoms of dehydration. A decrease in night vision, or dark adaptation, is also a common symptom. I've discussed this with experts, and they've cited as much as a 25% reduction in night vision with just a 1% drop in body weight by dehydration. I can't cite any specific studies that support this conclusion, but I've heard it enough that it sticks with me. That's not science but staying hydrated is well supported for any number of performance reasons. So one of the best ways that you can preserve and support your night vision is to stay hydrated. This should be a bit of a no-brainer at SFAS. Drinking water is a default position, but with all of the sweating that you'll do, even in the winter, you need to prioritize staying hydrated.

So the Army dedicates a measly 230 words to the topic, and they essentially conclude, "Darkness presents its own characteristics for land navigation because of limited or no visibility. However, the techniques and principles are the same as that used for day navigation." They're not wrong, but we think we've added some valuable insight to the body of knowledge. At the very least we added a couple extra thousand words to ponder. That's a gain of 10X!!! We'll close this discussion out with a reminder to "aim small, miss small." What we mean by this is that its well worth your time to take your time at night. Be as accurate as you can in all of the measurements and calculations. Don't introduce unnecessary error by rushing and being sloppy. Only go as fast as you can, without getting lost. Employ all of the same techniques that you do in your day navigation, just more deliberately. It'll be tougher to get 'un-lost' when you can't see anything. And last but not least, stay out of the draws. The *Draw Monster* feeds at night.

NO SHIT, THERE I WAS...JEFE, POR FAVOR, NO!

Night navigation can take on a whole new level when you do it in the jungle. Generally speaking, moving at night in the jungle is inadvisable. Snakes and animals and bugs and cliffs and danger abound. Surprisingly, some jungles aren't too hard to move through. The triple canopy prevents the bulk of the sunlight from reaching the jungle floor, so while the overhead cover is thick, the ground cover is fairly navigable. But, most of the time jungles are nearly impenetrable, day or night. I've spent a significant amount of time patrolling the jungles of Latin America and I've been in everything from mountain rain forest, to coastal wetlands, to sweltering triple canopy. The Mosquito Coast may be some of the nastiest stuff I've ever been in.

The Mosquito Coast is the eastern coastline of Nicaragua and the southern portion of Honduras. It's a generally low-lying area that boasts the most rainfall in Central America. It also boasts some of the densest populations of insects and jungle disease. It's quasi-remote and in a few places has some elevation that forms what can only be describe as the world's nastiest draws. It may be the primordial birthplace of the Draw Monster. It is thick. Have I mentioned that you should STAY OUT OF THE DRAWS!?!

This rugged terrain is made more challenging when you're moving with a group of less motivated travelers. I was moving through the area on a joint patrol with some Host Nation police and Army forces one sweltering afternoon. We had been dropped off inland by some helicopters and were making our way to the coastline. We were investigating reports of clandestine marijuana grow operations. Weed, it turns out, grows like a weed. You can spot patches from the air once you learn the telltale dark green color. The problem is that it's in the jungle and there is seldom a landing zone nearby. Some growers set up booby traps including improvised mines, punji pits, shotgun shell "toe poppers," and razors strung across trails at eye height.

Those sorts of countermeasures are rare for marijuana. They're not uncommon for cocaine processing sites along the Andean Ridge, so the tactic is known and often earns an outsized status relative to its actual threat. Patrols can get pretty skittish. Add to this the general hostility of the jungle and then plunge the entire situation into darkness. Let's just characterize the patrol as highly puckered. You may have this image of the local guys being one with the jungle and moving swiftly through the growth.

Living off the land and communing with nature. This was not the case. These guys were freaked out and we had a hit time to make on the coast.

Every time we heard a bird call or a monkey screech, they would all go to ground. Every time we hit a little intermittent stream, they are convinced that it's a giant anaconda (there are no anaconda in the Mosquito Coast). Every time they get snagged on a vine, they are convinced that it is a tripwire and death is imminent. So the going is slow and the men are nervous. This is what Green Berets are made for. Build rapport. Train skills. Inspire courage. Accomplish the mission. But not these guys, and not under these conditions.

After another unnecessary halt I pushed my way to the front of the formation. In my broken Spanish and no small amount of pointy-talky communication, I figure out that the point man is convinced that a jaguar is poised ahead, and it would be best if we just stopped for the night. This is our ninth such 'tactical pause' in the past few hours. I am growing impatient, and the clock is ticking. It was dark. It was thick jungle. It was hot and nasty. We were smoked. But I'm certain there was no jaguar (or anaconda or Draw Monster). The patrol leadership is adamant that we hunker down. I try all of my negotiating techniques, but the patrol has decided by default. They implore me, "Jefe, Por favor, NO!" It's always best to never give an order that you know won't be followed.

So we hunker down and set up a security perimeter and let the jungle win. In the morning, now rested and less scared, I rally the troops for the final push to the coast. I figured that it was important that I set the pace for a quick movement. No excuses today. So I go up front and lead the point man out. We walk a few dozen meters and come up to the spot where he was convinced the jaguar was waiting to ambush us. What did we find? A stump. A fucking stump. No danger, just a rotting stump of on old tree. I try to remember this when I'm out navigating at night. Rarely is there a jaguar or an anaconda. The Draw Monster is real, but less dangerous than you might imagine. You are in charge, and you must move at night. You are the Jefe, so don't take no for an answer.

Many SFAS candidates think they can hunker down at night and just move quicker during the day to make up for lost time. This will not work. Almost half of your allotted test time is under the cover of darkness. You must learn to navigate in the dark. Be the Jefe.

9

ALWAYS LOOK COOL

"The ability to be cool, confident, and decisive in crisis is not an inherited characteristic but is the direct result of how well the individual has prepared himself for the battle."
– Richard M. Nixon

The observant reader may note that the title of this book is one of the 3 Rules, and now this chapter follows suit. I thought that this was a nice nod to the Special Forces heritage and I'm always keen to stay connected to the bigger picture. Call it strategic thinking, but I've always found great utility in the 3 Rules. They are so much deeper and more useful than just a pithy commentary on the Special Forces lifestyle. As I explained in some detail in *Ruck Up Or Shut Up*, the 3 Rules have both a literal and a symbolic definition. Looking cool isn't just about your physical appearance, although that certainly has a significant impact. Looking cool is also about being in control, demonstrating competence, and presenting the best version of yourself. Nobody wants to follow a spaz, so get your shit together.

This final chapter is our catch-all chapter. We will cover all of stuff that we haven't covered yet but we still need to document as part of your land navigation learning experience. Some of it will make you look cool. Most of it will help you to be competent. Some of it is just stuff that I have to rant about that is tangentially aligned with land navigation, and there's no other place to put it. This doesn't mean that this stuff is any less important. It just means that it's tougher to easily categorize.

Terrain Association

There are broadly speaking two methods of land navigation, terrain association and dead reckoning. They each have their pros and cons and the best navigators use both on a case-by-case basis. We have already introduced the concept of terrain association earlier when we discussed route selection. You will recall that we chose checkpoints (CPs) in part based on their ease of recognition. The 6-way intersection that we've used for many of our demonstrations is a great example. Imagine if you were physically standing at the 6-way intersection. You had a map in your hand and a compass dummy corded to your belt. You could use your compass to determine which direction was north and then you could orient your map, so it was aligned correctly with the top of the map pointing north as well. This map orientation is a simple but critical skill that you should always learn to incorporate.

At any rate, you are standing at the intersection with your map appropriately oriented. Those roads are pretty distinct, aren't they? They are fairly prominent, and they follow a regular and uniform trajectory as they intersect. How easily could you orient yourself to this location? One of the roads is running almost perfectly north-south. If you were standing at that intersection with that map and compass, do you think that you could figure out which road was which? Could you associate which road on the map corresponded with which road on the actual ground? My guess is that you wouldn't struggle to do this. The features are very prominent and there isn't much to obscure their orientation. What you have just done is *terrain association*. You have associated the terrain on the ground with the terrain represented on the map. Terrain association.

Now, "that's pretty simple," you might argue. The roads are easily identified. They follow a regular and uniform orientation. They are easy to see on both the map and the ground...they're roads for Pete's sake! Any fool could see the intersection and easily associate it with the map. But it is that simple. That's terrain association. Of course, not all terrain is so recognizable. It's not uniform, it's not man-made, and it's not perfectly oriented. But it is on the ground and usually on the map. You just need to learn to recognize it. Think back again to our route selection discussion. We used that 6-way intersection as a checkpoint and continued south to our next check point, which was the east-west running bowling alley. We picked that as a CP because it was easily recognizable, right? We picked that as a CP because it was easy to associate that terrain with what was on the map. We would know when we walked up to the bowling alley...because it would be a big ass open linear field that was oriented east-west. We could associate that terrain with the map. We would know where we were. We would be navigating, using terrain association.

This was an ideal route because we didn't need to shoot a precise azimuth. We just had to move generally south from the 6-way intersection. We were using the north-south road as a guide, a handrail. We were moving along that handrail until we got to the next terrain feature, the bowling alley. We were using all of these little bits of information to help us negotiate the movement. We were terrain associating. Some terrain doesn't afford us the option. Flat, featureless terrain like an open desert doesn't allow for terrain association. Some terrain is full of features. As we discussed in the *intersection, resection, and modified resection* portion in Chapter 6, the entire concept is predicated on associating terrain. You are looking off in the distance and identifying a hilltop and then shooting an azimuth to it. You've just done terrain association.

It's easy to grasp this concept in the sterile and comfortable pages of this book. We almost stumbled into it just now. You were doing terrain association way back in the early chapters, you just didn't know it. But there's no reason why you can't do this on the regular. You will need to get comfortable in the woods. You'll need to develop the sense of your surroundings and to recognize terrain, even less significant terrain like gently rolling contours (there we go back to chapter 4). You'll need to learn to keep the map oriented and keep a good pace count. You'll need to learn to manage your cognitive load and do your RV procedures and follow the rules. But there's nothing "advanced" about the technique. It just takes a little bit of practice. You have to train. But that's all *terrain association* is. Associate the terrain you see on the ground with the terrain you see on the map.

Dead Reckoning

The second method of land navigation is dead reckoning. In contrast to terrain association, this technique is entirely predicated on precise distance and direction, as dictated by the compass and pace count. These calculations are independent of the terrain and the map. Inexplicably, some trainers start teaching land navigation with dead reckoning. I think this is a mistake as it reinforces the notion that terrain association is some advanced witchcraft. In reality, terrain association is much more natural than dead reckoning. It's just a little more nebulous and takes some discernment. Dead reckoning is science, terrain association is art. A benefit of dead reckoning is that as long as you have the tools, a way to measure distance and direction, then you can navigate. You don't need the terrain. You can follow an azimuth on your compass and count your paces with a bucket on your head. Let's see how we practice this science, bucket-head.

As we discussed in Chapter 3, determining the distance on a map is pretty simple, particularly a straight-line distance. All you need is the start point and end point, a straight edged piece of paper, and access to the bar graph. As we discussed in Chapter 2, plotting an azimuth on the map is pretty simple as well. All you need is the start point and end point, a straight line connecting the two, and access to a protractor. So determining the two data points, distance and direction, is already a skill that you have developed. Now we're just applying it in a new context.

> Have you noticed how we keep referring back to previous chapters? Have you noticed that we repeatedly circle back to concepts that we already introduced and are now applying them to different situations? Are you beginning to understand that while land navigation isn't really easy, it is pretty simple? Know where you are (your start point), know where you are going (your end point), know how you are going to get there (route selection). There is a ton of nuance and skill in the route plotting and actual execution, but it's all very doable. There really is nothing advanced about land navigation. Back to the distance and direction!

Now that we've obtained our desired distance

and direction from our map calculations we need to "program" this information into our compass. We'll cover pace count next for distance, so lets just focus on direction. We plotted our azimuth on the map, and we did the appropriate G-M Angle declination calculations (when converting from grid azimuth to magnetic azimuth you add the G-M angle...you don't have to remember this, it's written on the bottom of the map). Now we have our azimuth. We 'program' this into our compass by looking down on the compass...hold on, let's talk about the compass some more.

Compass

As we already discussed in the night navigation portion in Chapter 8, there is some nomenclature we should cover. What do we call the various parts of the compass? Ideally, you would have a compass in your hand, and you could follow along, but this diagram is fine for our purposes. Also, there is a reason why we cover the compass in a few places in this book. I want this lesson to be crystal clear in your mind. I don't want there to be any doubt that we are absolutely relying on a compass and that compass is the Cammenga 3H. Don't go all fucking Hollywood on me here. Yes, I am a Green Beret. Yes, you want to be a Green Beret too. But you are not Rambo (even Rambo had a compass on his survival knife, by the way). You are not a secret agent man, and you will not have to improvise a compass. You will not navigate by the stars or unlock your inner medicine-man shaman and tap into ancient powers. You are using a compass. You are using a proper compass. A Cammenga 3H. That's what a Commando uses. Here endeth the lesson.

A proper Cammenga 3H, note the tritium markings.

A Cammenga Model 27, close but no cigar.

The Cammenga does have some nuances. It is a tool, and it is only as good as the craftsman that wields it. So let's discuss it a bit more. The graphic gives you all of the nomenclature, but I want to cover some of how it works so you can learn to inspect, maintain, and use it appropriately. Make certain that the hinge is tight. It should move freely, but not floppily. When it's open is should remain open, when its closed it should remain closed, and when it's partially open it should remain partially open. Make sure that the *luminous sighting dots* (in the *cover*) are intact and bright. Similarly, make certain that the other luminous elements are intact and bright. The *sighting dots*, the *short luminous line* on the *bezel ring* (aka rotating bezel), and the *luminous magnetic arrow* (little tick mark on the north seeking arrow) are all self-illuminating tritium. They do NOT need to be charged by artificial light to be seen. The other luminous elements are the luminous patches that are underneath the E (east) and W (west) symbols on the *floating dial* and the patch under the *fixed index line* (which aligns with the *luminous sighting dots* in the cover). These three patches are NOT self-illuminating tritium and should need charging with artificial light periodically. A point of inspection is how bright they are after charging and how long the brightness lasts. Patches that do not charge well or that quickly lose their charge might be cause for replacement.

Make certain that the *bezel ring* rotates, but not freely. It shouldn't bind at all, but it should resist movement to the point that you can feel each "click" as you rotate it. Each click indicates 3-degrees of movement, so it should be 120 clicks for a full revolution. Make certain that the *sighting wire* (in the *cover*) is intact and that it aligns well with the *sighting slot* (the notch in the *lens or rear sight*). The lens or rear sight apparatus needs some inspection. Make certain that the *slighting slot* is intact, and the *lens* is clear. The *rear sight* apparatus serves both the obvious function of sighting the *wire* and observing the azimuth on the *floating dial*, but it also locks the *floating dial* when the *rear sight* is "closed" or pressed into the *bezel ring*. This prevents the floating dial from spinning, which is good because it reduces wear and misalignment on the *floating dial*, but if you're not observant the *rear lens* might be closed just enough to prevent the *floating dial* from spinning freely and not be obvious that it is obstructing it.

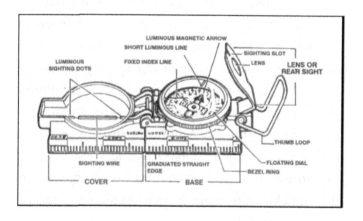

That's enough nomenclature, let's get back to actually using it. We 'program' this stuff into our compass by looking down on the compass and reading the azimuth by aligning the *floating dial* numbers with the *fixed index line*. In doing this you are now 'pointing' the compass in the desired direction. But there is some nuance to this. There are actually two "official" techniques when pointing the compass, compass-to-cheek (which everyone calls cheek-to-compass) and centerhold. To use the compass-to-cheek technique you half-close the *cover* of the compass containing the *sighting wire* to a vertical position. You then fold the *rear sight* slightly forward and look through the *rear-sight slot* and align the *sighting wire* with a desired object (a tree, a rock outcropping, a building) in the distance. Then look down at the dial through the eye lens to read the azimuth.

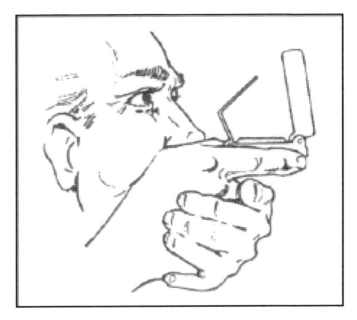

Cheek-to-compass

The second 'official' technique is the centerhold method. To use this technique, you open the compass completely so that the *cover* forms a straightedge with the *base*. Move the *lens* (*rear sight*) to the rearmost position (fully opened), allowing the *rotating dial* to float freely. Next, place your thumb through the thumb loop, form a steady base with your third and fourth fingers, and extend your index finger along the side of the compass. Place the thumb of the other hand between the *lens (rear sight)* and the bezel ring; extend the index finger along the remaining side of the compass, and the remaining fingers around the fingers of the other hand. Pull your elbows firmly into your sides; this will place the compass between your chin and your belt. To measure an azimuth, simply turn your entire body toward the object, pointing the compass cover directly at the object. Once you are pointing at the object, look down and read the azimuth from beneath the *fixed black index line*.

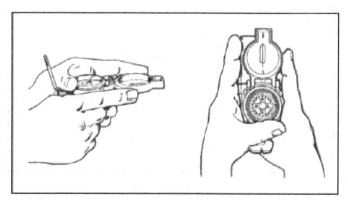

Centerhold

Cheek-to-compass is more accurate, but it can take a lot of time to get a reading. Centerhold is faster to get a reading, but it can be slightly less accurate. You have to balance the need for precision with the need for speed. The fastest way to get a direction is terrain association. So your job becomes planning routes that can take advantage of the techniques that afford you the correct amount of speed, precision, and control. Terrain associate to make good time, navigate to an easily identified control point (checkpoint/attack point), and then dead reckon the last leg for maximum precision. If you use this methodology, you will likely need the third 'unofficial' technique of compass employment which is just whip it out and get a quick orientation. "That way is north and now I can confirm that I am moving along the correct ridgeline to my next checkpoint." This is how we maintain awareness of our direction.

Pace Count

The way that we maintain awareness of our distance is with a pace count. We need a way to keep track of how far we have walked so we can track our progress across the terrain. This many steps equals this much distance; we call this our pace count. To determine our pace count we lay out a course with a fixed measurement and walk that course and count our steps. What this looks like is usually a 100-meter course. You can determine this measurement with a GPS, a measuring wheel, an app, or a predetermined course. It might be a tough task finding a predetermined course, but running tracks can be helpful. The proper length of the first lane of a competitive running track is 400 meters, so you could walk that lane and divide by 4 to get your pace. Some tracks are not built to this specification, instead being a legacy to imperial distances such as 440 yd, but the difference is 2.3 meters. So, in the aggregate, you could get close enough. Walk the track in lane 1, divide by 4. That's your pace.

It's important to note that your pace is not your steps. If you count your steps, you are counting every time a foot hits the ground. If you are counting your pace, you are only counting every other step. Most guys count every time their right foot hits the ground. We do this because in military drill and ceremony, you always step off with the left foot first. So this is just ingrained in our psyches, and it carries over into other life tasks. Counting your paces versus counting your steps is a good way of managing cognitive load, which we'll talk about in a moment. You are literally cutting your cognitive load in half for this task by counting paces versus steps.

But determining your pace count on a predetermined course is a bit canned. Yes, this is your pace count on a track (for example), but you won't be navigating on a track. You will be navigating in the woods. So there is some merit in determining your pace count under various conditions. Heavily vegetated and sloped terrain can impact your count. So can carrying a load, like a rucksack. As discussed in Chapter 8, navigating at night can impact your pace count. Most guys take shorter steps in the dark out of a natural caution. You should also determine your running pace count. And finally, we should recognize that your pace count may change when you get fatigued or injured. All of these factors may only change your 100-meter pace count by 2 or 3 paces, but over a 6,000-meter movement this could result in a significant difference.

Pace beads then become a tool to help you keep track of all of this counting. Pace beads are usually a hank of paracord with some beads threaded on to it. There are 4 beads in the top portion and 9 beads in the bottom. The overall length is user preference, but super long beads look goofy and present a snagging hazard. Super short beads might not maintain enough positional distance between up and down and thus be inaccurate. Usually about 12 inches finished length is the sweet spot. As you walk along and count your paces, every time you walk 100 meters you slide a bead on the bottom portion down (start with all beads in the up position). As you go, you'll eventually get to 900 meters (9 beads slid down now) and you'll have walked another 100 meters, so 1,000 meters total. At this point you slide one of the top 4 beads down and you will reset the 9 bottom beads back up. So one top bead down denotes 1,000 meters. You simply repeat this process as you go, sliding a bead for every 100 meters. As such, if you had 2 top beads down and 3 bottom beads down, this denotes 2,300 meters. You don't have to keep track of how far you've gone because the pace cord does that for you. You just have to keep pace, slide a bead every 100 meters, and the cord keeps track of your distance travelled.

Pace Beads, aka Pace Cord

You should also keep time to keep track of how far you have walked. This method works by timing how long it takes you to move 100 meters, say 1 minute (my time is usually 1 minute and 13 seconds, but let's use 1 minute for ease of demonstration). Therefore, if it takes me 1 minute to move 100 meters, it will take me 10 minutes to move 1000 meters. Now I have a distance estimation tool that is independent of my pace count. As long as I make note of when I started my movement, I will have a reasonable approximation of the distance I have traveled. Again, this time may vary at night, under load, under duress, or on varying terrain, but it is another valuable data point to help us stay oriented along our route.

So let's recall back to Chapter 7 with route planning. This whole process begins with knowing where you are (your start point) and knowing where you are going (your end point). The process for connecting these two points is planning your route, your distance and direction. You now have two methods for determining distance...your pace count and the time. You have two methods for determining your direction, azimuth and terrain association. Your route can be as simple as a straight-line azimuth for the determined distance (dead reckoning) or an actual plotted route that accounts for obstacles, terrain, checkpoints, attack points, handrails, boundaries, backstops, etc. This is a total of four data points.

Distance

- Pace Count
- Time

Direction

- Azimuth
- Terrain Association/Orientation

If you didn't keep track of time and you only kept track of your pace, if something happens to your pace count then you now have zero data points for distance. But if you keep track of time and you lose your pace count, you still have a distance data point. And time exists whether you are keeping track or not, so you might as well use it to your advantage.

Another way to think about this entire process, terrain association, azimuths, pace count, and time, is by plotting it out on a map. Let's assume a worst case set of conditions. You only have a few pieces of information. You know where your started and you know how long you've been walking. Remember, time exists, whether you acknowledge it or not, so you'd might as well start acknowledging it. With these two data points – absent of any azimuth, terrain association, or pace count – you still have a fair idea of where you are. If you know where you started, and you've been walking for twenty minutes (assuming your pace time is 10 minutes every 1000 meters), then you can draw circle at 2000 meters with your start point in the middle, as such:

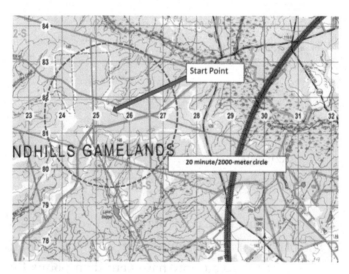

20 minutes, 2,000 meters...I'm somewhere in the circle.

So logically, knowing nothing but your start point and the 20 minute/2K meter threshold, you are somewhere inside this circle. So you're "lost", but you're not *completely lost*. You just have to refine your position within this circle. This is actually pretty simple. Follow along. You can eliminate half of this circle just by identifying what direction that you didn't go in. You don't even need an azimuth, per se. You just need to know that you didn't go north. How hard is that? You knew you were in the start position and you starting walking south. Not "187 degrees" specifically, just generally south. Not north. You can immediately eliminate half of the unknown now, as such:

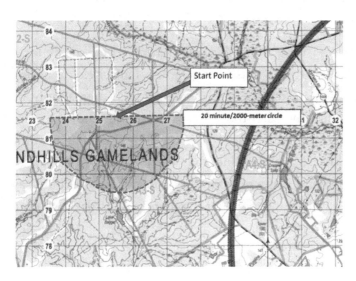

I didn't go north, so I'm somewhere in the southern half of the circle.

I know, through this logical sequential process, that I am somewhere in this half circle. And unless I wandered violently, I am likely somewhere at the edge of the circle. I could narrow down my "lost zone" quite a bit, as such:

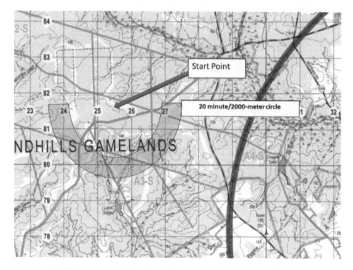

I didn't wander aimlessly, so I'm somewhere at the edge of the circle.

I have now effectively narrowed down my location from a massive circle to a half circle, to an arc that is 20% the size of the original area simply by knowing the time I was traveling and the very general direction I was traveling in. Imagine now that I could refine that even more by instead of having just a general direction of south, I could say southeast. I could cut the margin of error down even more, as such:

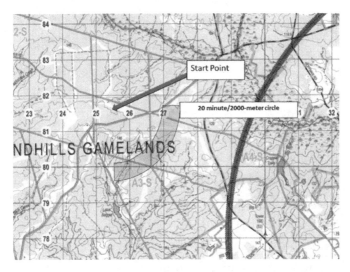

I'm kinda southeast, now I'm narrowing it down! I'm just using time and common sense.

And the process can continue as you refine all of the available data - time, distance, direction, azimuth, identifying terrain (terrain association), checkpoints, handrails, boundaries, backstops, etc until you can adequately determine where you are. That's really all land navigation is.

Search Procedures and Techniques

Once you reach your attack point (the last checkpoint before your final point – this is usually where you should start dead reckoning) you should take a deliberate pause. Not necessarily a lengthy pause, but a deliberate one. I recommend a special set if RV procedures (remember Chapter 7?) to be conducted at your attack point. Your attack point should be a very recognizable feature that you can definitively identify both on the map and on the ground. This point serves as a conclusive jumping off point for the final leg of your route. I always pause and get myself sorted out, physically and mentally. I take a few moments and hydrate and take in some calories. I want my brain working at peak performance. I'm about to have a likely engagement with a cadre member or at the very least find a point. I blow my nose and wipe my face clean of all of the detritus that I collected on my grueling journey. If I look haggard and vexed, then I will be assessed as such. I ensure that I have all of my gear. I double check that I have my route data and my scorecard handy. I might even rehearse what I will say to the cadre so that I'm not stumbling over my words. But most importantly, I'm preparing my search techniques.

As I'm closing in on my end point from my attack point, I am elevating my alert level. I'm actively scanning. I'm looking for the point itself and potentially for other candidates or even cadre that are moving in and around the point. Movement catches the eye, so use this final approach to see that movement. When I get to the end of my final leg, and I haven't discovered my point I always pause. Just pause for a moment, stop moving, and take in your surroundings. When I pause, I always look to my left to start my search. You may remember in Chapter 8 during our *Trust Your Sensors* discussion, we discussed drift. My natural drift is to the right (I am right hand dominant), so I always look to the left for my point. My final leg, from my attack point to here, isn't very long so I shouldn't have too much drift. But my habit is to still account for this drift and look to my left. If the end point isn't apparent, then I start my search procedures.

I always start my search procedures by conducting SLLS (pronounced sills). SLLS is a patrolling procedure that stands for Stop, Look, Listen, and Smell. I stop so that I can more readily identify movement, as discussed earlier. I look for obvious movement, signs of trail, and matching terrain. I listen for any activity that may alert me to the end point location. Manned points, like during the STAR exam, may have vehicle traffic or you may hear others at the point. Finally, I smell. Some manned points will have small fires and after a few days in the woods you can smell food surprisingly clearly. But you are only likely to notice these small clues if you take a moment to gather yourself and deliberately seek them, even as a passive scanner.

If I haven't discovered any clues from my SLLS check, then I can begin my search patterns. All of the search patterns do the same thing. They allow the searcher to systematically cover all of the desired terrain in a deliberate and comprehensive manner. The shape of the pattern doesn't matter, just the end state. The most common patterns are cloverleaf, box, and concentric circles. The common thread is that they all establish a fixed center point and then branch out from there to cover the desired area. All the while that you are walking the determined pattern you are actively scanning. Once you reach the apex of the search pattern, or at regular intervals (depending on the pattern type that you choose) make certain to stop and conduct SLLS. Regular, deliberate, and comprehensive search patterns will give you the confidence that you have adequately covered the search area. Assume that the point is there, somewhere, and only after exhausting your search patterns and SLLS stops, maybe 30-45 minutes depending on conditions, should you return to the attack point if you can't find your point.

In the event that you can't find your point, return to your last known position, your attack point, and confirm your plotting. I can't tell you how many candidates, in a rush to start their route, mis-plotted the point on their map. It's easy to be an entire grid square off by simply transposing one number. You're under stress, you're under extreme cognitive load, and the very first thing that degrades when placed under a time crunch is your ability to plan. So mis-plotting is a not uncommon error. Best to take your time early on. Mistakes tend to cascade, they start small and snowball into bigger issues. Control what you can when you can control it, so when stuff you can't control starts happening, you are in the most advantageous position possible.

Map cases (and Protractors...and pens and pencils and notebooks...)

Map Cases

I spend waaaay too much time thinking, obsessing really, over mundane gear choices. But all you have to do is go back and re-read *No Shit, There I Was...The*

Case for Map Cases to understand that this stuff isn't important until it is important. And then it's critically important. So, if this gear stuff isn't important to you then just skip ahead but let this serve as a warning...I told you so. Think about it now, when you're warm and dry, while you're well rested and fed, and while you're not hopelessly lost. I could tell countless stories where my pedantic obsession with gear saved my ass, quite literally saved me from having my ass hang out in the wind on full public display. So scoff all that you like, but you will do so with your ass exposed and that's a violation of the 3 Rules.

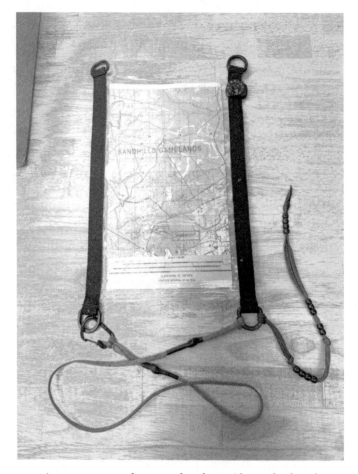

A proper map case...dummy cord on closure side, pace beads, and a clipper compass.

Let's start with map cases. This used to be a no brainer as there was really only one choice, the Seal-Line series was the Gold Standard. What you were looking for was a rugged case that could be folded and tucked away repeatedly without cracking, was waterproof...legitimately waterproof, not weatherproof or water resistant, was markable and erasable, had a positive closure, and had secure lashdown points. In recent years, thanks to environmental concerns, most manufacturers have transitioned from a PVC material to a TPU formula. This includes the scoundrels at SealLine. In doing so they have abandoned their customer base and committed treason against functionality (yes, I'm being dramatic...but not without standing). This new TPU formula does create a supple clear material that is quite strong, and it is waterproof. But when you use a standard alcohol marker on it, it permanently marks the plastic and cannot be erased. A massive failure. I have spent months researching, torture testing, and comparing every conceivable map case that I could get my hands on. The truth of the matter is that the pickings are slim. You either have to choose a waterproof case that stains, or a leaky case that remains stain free. Why some manufacturers insist on using a zipper closure is beyond me and Velcro is only slightly better, but only when appropriately applied in conjunction with a folding closure. In the interest of not clogging up this book with too much dated material, I'll just point you to TFVooDoo.com where we have an article (along with many gear recommendations and reference material) that we will keep continually updated. In this Map Case article, we will maintain the most current list of tested map cases, our recommendations (along with links to the actual item...not similar items, but the version, model and material that we endorse), and any other ancillary information. Is it overkill? Yes. Is it pedantic? Also Yes. Is it worthy of the effort? Absolutely yes.

Whatever model of map case that you choose, we

recommend configuring it in a specific way. First, put a dummy cord on it. Make the dummy cord long enough to loop over your head (so that the map case can hang freely on your chest for quick reference), but not so long that it easily becomes entangled. Make certain that this dummy cord is affixed to the side of the case that has the opening on it. What you are looking to do is configure your dummy corded case so that when it hangs freely from your neck, the closure is at the top, thus eliminating or at least reducing the chance of the stuff inside spilling out. If you've ever lost a map or a scorecard because you didn't notice that the case wasn't securely closed and everything slid out during movement, then this is a mistake you will only make once. You don't have to make this mistake at all because you configured your map case correctly, like I just told you. Finally, you should affix your pace beads to your map case. This might not work for patrolling as it requires you to have your map case out to use your beads, but for land navigation training this is ideal. What this configuration does is that it forces you to put hands on your map case every 100 meters. Every 100 meters you reach up and slide another bead down, so every 100 meters you ensure that you have your map case. If you configured it correctly (dummy cord on the side of the closure) you also ensure that you have your map, your scorecard, and your route notes (Chapter 7 -> Route Selection -> Write it down). So a simple little thing like picking the proper map case, properly configuring it, and properly accessorizing it, helps to ensure that you never get lost. Gear matters.

Protractors

You will be issued a protractor with limited facilities to modify it. So in many ways you get what you get. But if you get the opportunity to modify your protractor there are a few things that you can do to optimize it. First, check to make certain that the die cut is accurate. Often times, when the protractor is initially cut at the factory, it is misaligned. On the perimeter of the protractor, where you read the degrees, this isn't really an issue as long as you can see the markings. But the coordinate scale cutouts, the little triangles (1:50,000 is really the only one you're worried about) are important. If those cutouts aren't accurate, they can really throw your plotting off. If you do trim them, make certain that you are being accurate and precise. If you want to trim the mils scale from the perimeter, make certain that it is straight. It won't make any difference when measuring the azimuth, but if you use this edge to draw azimuths on the map, a poorly cut protractor can make accurate plotting troublesome. The final thing to check is the crosshairs. Sometimes the protractor will come with a hole already punched at the center of the crosshairs, sometimes you need to make that hole yourself. If its already punched, make certain that it is exactly in the center. If you make it yourself, be precise. The best way to make this hole is with a hot needle and make it as small as possible. The task here is to thread a piece of string through this hole and use this string to measure your azimuth. When you place the crosshairs on your origin location and pull the string taut, it will be in line with the azimuth that you need to measure. Many guys will erroneously make a big knot on the end of the string to keep it from pulling through the hole, but this often introduces some error as that knot is so large that it makes placing the crosshairs precisely problematic. I recommend tying the string in a continuous loop, thus eliminating the knot problem. I use gutted 550 paracord innards, and many types have multicolored inner strands. Pick a contrasting color to make reading the azimuth easier under varying light conditions. There is your protractor sorted.

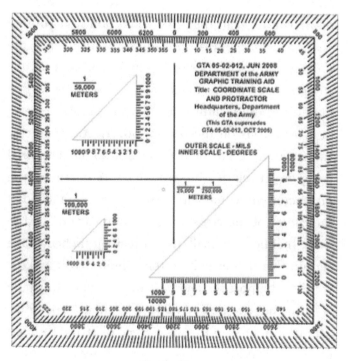

Standard Issue

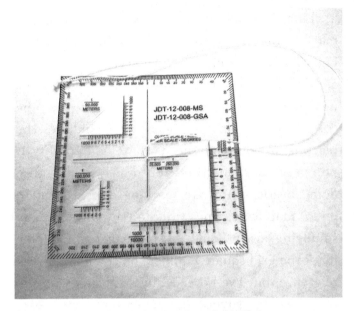

Commercial Variant, Properly Modified.

Pens and Pencils

What are you, a nerd? Why do you care about pens and pencils? Who cares? Just pick the stuff they issue from the supply cage! If it's issued then it must be good, or at least good enough! Shut up. Shut your mouth that is full of bad takes and horrible advice. We just spent several pages masticating over map cases. Do you think we are going to dedicate nearly 1,000 words to map cases and just ignore writing implements? Not a chance. Now, I'll admit that there is a pen and pencil subculture that is...excessive. But we need that stuff. We need guys who go *too deep*. I've tapped into that hive mind, and I appreciate the particular part of the spectrum that they occupy. I used them to shape my advice here.

First, let's talk alcohol markers. These are the markers that we referenced in the map case staining. There is only one choice, the venerable *Staedtler Lumocolor* fine point (or super fine) permanent marker in black. You can certainly use other colors for marking obstacles or boundaries, but black is all you really need. Don't bother with the non-permanent markers. You aren't a staff officer doing overlays. You are marking critical locations on a map case, and you don't want that stuff smudging away. Some folks like a highlighter to mark the grid square numbers on the map for easy reference, but I've never really struggled with this, so I find this superfluous. User preference really. You can use these Staedtler markers for a regular writing instrument, but that blunts the tip prematurely. We have some thoughts, shocker, on regular pens.

Pens are suboptimal. There, I said it. I can hear the autists screaming. I get it. I'm a pen enthusiast myself. Gel pens, fountain pens, Japanese pens (if you're in the game, these are the gold standard). I carried a Lamy Safari fountain pen for years. It makes everything you write down automatically 10% smarter. But they take some maintenance, the ink smudges, and I'm leery to drop more than a couple of bucks on an item that is essentially (especially in the context of land navigation) a disposable item. If you are going to carry a pen, I recommend a pressurized pen, like a Fisher space pen or the Rite-in-the-Rain pen. Non-pressurized pens do not write reliably on wet paper, and you must assume that your notebook will get wet at some point. So let's talk about notebooks.

This one really is a no-brainer. There is only one notebook that you should use for field use – Rite-in-the-Rain. That's it. No more discussion required. This is the universal gold standard. I love my Field Notes and my Moleskines, but they aren't appropriate for the hard use scenario of land navigation. You need waterproof paper. Rite-in-the-Rain is the best choice. The SFAS packing list authorizes only one notebook, up to 4 x 6 inches. So get yourself a 4 x 6 inches Rite-in-the-Rain notebook. You may also note that the inside of the Rite-in-the-Rain notebook has the bar scale printed on it. This can be very useful when calculating distance. And the stiff plastic cover serves as a useful straight edge for plotting accurate azimuths. Rite-in-the-Rain, that's the pick.

Lastly, pencils. Yes, pencils. Pencils are superior to pens. They don't freeze, they're less likely to smudge, and they write on wet paper. Pencils also have the huge advantage of being erasable, unlike most pens (erasable ink is inferior, ignore it). So pencils are the way to go. Mechanical pencils specifically. I understand the attractiveness of the simple wooden pencil. Easy to sharpen, no parts to break, and cost efficient. But a good mechanical pencil is cheap, always sharp, and infinitely superior to pens. Always choose a .5mm (the other common size is .7mm) as the extra precision is critical. The other nice thing about mechanical pencils is that you can get them in nearly any color of barrel (not lead...use regular lead, don't be a weirdo). I always choose a bright color, so I am primed to see it if I drop it or am otherwise searching for one. This leads to my final point and segways perfectly to the last point of the chapter, *cognitive load*.

I spend so much time obsessing over these little details for a reason. Primarily, I want good gear that won't fail when I need it most. I hate cheap gear and I've learned that the 'buy once, cry once' theory of spending more money once is better than spending less money more often is well supported. So I research deeply before I commit to a purchase, even "cheap" stuff. The secondary reason, counterintuitively, is that I spend time thinking about this stuff so I don't have to think about it. I don't want to be thinking about my map case, or my pens, or my paper when I'm out doing land navigation. I want to be able to use my brain to think about the terrain and to assess my routes and stay engaged with what I'm doing in the moment. I want to manage my cognitive load.

Managing Cognitive Load

If you can't think straight, then you can't act straight. At SFAS you are being judged on both, simultaneously. You need to develop systems and processes that allow you to structure critical repetitive tasks so you can stop actively thinking about them and free up that cognitive capacity for real time processing. Because land navigation is concurrent cognitive and physical load, it is a prime opportunity to use this technique. Let's use *RV procedures* as an example.

We previously discussed RVs as interchangeable with CPs and attack points and this heuristic is still valid. You should execute this RV procedure at regular intervals along your route, so CPs, RVs, and APs are all compatible. Essentially, you should develop a set process that you do every time you come upon an RV, CP, or AP. If you execute this procedure regularly then you eliminate a significant cognitive load. Every SFAS class, many students lose maps, scorecards, end even weapons. Almost every student loses their orientation at some point. By establishing simple RV procedures, and being disciplined about executing them, you can turn this liability into an asset.

If you develop an RV procedures system, you eliminate the risk of misplacing key equipment. When I follow my RV procedures, I am certain that I never lose my equipment. If, for some reason, I have lost some equipment I can limit my search area to between where I am and the last position that I conducted my RV procedures. If I planned my routes correctly and I put a waypoint every 700 to 1000 meters, then I can focus my search in that specific area, instead of the entire route that I have walked. I conduct these RV procedures before I leave my start point, at every check point, at every prominent navigational aid that I encounter (even if it wasn't pre-planned), and at my attack point just prior to arriving at my end point.

My RV procedures are just an intuitive part of my

navigating and because I don't have to worry about my gear, I can use that cognitive capacity for actively navigating. When I am closing in on my end point and I cue myself, I am mentally preparing to engage in a very critical information exchange with a cadre member or a point sitter. I am preparing myself to be assessed. I've made certain that I have all of my equipment and I have sorted myself out to make certain that I am presenting the most competent version of myself possible. When I have that information exchange, I can respond with a clear mind and I can take decisive action. I gain this huge advantage by simply establishing a habitual RV procedure. I am managing my cognitive load.

You should apply this thought process to many of the functions you will routinely engage in at SFAS. Every time I put my ruck on, I grab it by the frame and lift it up over my head and slide into the shoulder straps. I do this not just because it looks cool, but because it helps me check to make certain that my ruck is packed correctly. If I was digging in my ruck and forgot to secure it correctly or have some loose straps, this procedure will reveal these deficiencies to me. This system served me well throughout SFAS. When I am nearing the end of my land navigation route and I am searching for my point, I always look to my left first. I am right hand dominant and therefore I naturally drift to the right while walking. Logically, my point should be on my left. I don't have to think about it, I just do it. It's decisive action and it took zero deliberate thought. If you develop enough of these heuristics, then you can free up significant cognitive capacity. A massive advantage.

Restrictive Terrain Procedures

This could really be called "how not to get your ass handed to you in a draw", but Restrictive Terrain Procedures sounds more marketable. This is an extension of managing cognitive load and is a way to give you some agency over the process. No matter how many times I say, "Stay out of the draws!", the reality is that you are almost certainly going to have to negotiate a draw or two at some point. So I want to give you some tools do so with as little pain as possible. First, be deliberate in your navigation. Don't just wander into a bad situation. Evaluate all of your options, develop the best route possible, and avoid as many obstacles as you can. If you determine that you must cross a draw, then this simple mindset shift gives you some control. You are still going to get punched in the face, but at the very least you get to decide which cheek is getting smashed. This mindset matters more than you might think.

When you approach the draw, while you are still on the high ground prior to walking into the mud and thick vegetation, start looked for the best way through. Often times you won't be ablate really identify and passage like a trail. But perhaps you can see the far side the draw and that can help you determine the shortest route through. Look for brown or dry patches both in the draw itself and on the far side. You can't find a good way in, but you can perhaps start plotting a good way out while you have a good vantage point. As you are descending the slope you should start preparing yourself physically by making certain that your pockets are secured, your map is tucked away, and you have your gloves on.

As you approach the edge of the draw, generally marked by the thickening of the vegetation, attempt to identify the path of least resistance. There is almost always a game trail present. The deer are moving through there somehow. It might not be a complete trail with unrestricted passage, but even a slightly beaten route is better than no route at all. Before you plunge in be certain to conduct RV procedures again. This way you will know for certain if you lost anything in the draw, or before you got beaten up by the Draw Monster.

Once in the battle, be deliberate in your movement. Stay directionally oriented; you don't need to follow an azimuth per se, but don't get turned around. Don't freak out if you get stuck. Extract yourself. Be calm. Always look cool. Be El Jefe. You might have to back out and renegotiate, but be deliberate. Be mindful of the terrain that you spotted on the other side of the draw as you were descending the slope. That's your target. Go to it. There's no need to stop and do a map check in the draw. Stop doing that. You are wasting time and you have that map tucked away for a reason. The terrain didn't change and you didn't move far enough to warrant a map check. Keep moving.

Once you emerge from the draw, immediately do

another RV procedure to ensure that you have all your gear. A pro tip is to also do an RV procedure at the apparent midpoint of the draw crossing. This is usually marked by a waterline of some sort. The more that you can isolate potential 'lost gear search areas' the better. This is both a physical advantage and a psychological one. The more control, even artificial control, that you can apply to this battle you are in the better. Yes, you are taking haymakers to the dome from the Draw Monster, but at least you still have all of your gear! This mindset is critical.

So, that's the cool stuff that we couldn't jam into the other chapters. Some of it is land navigation specific, some of it is universal, and some of it is full-blown weirdo. But, I include it all because at some point or another during some of the thousands of miles that have navigated, patrolled, and wandered the planet it has saved my ass. Good judgment comes from experience, and good experiences often come from bad judgement. These are lessons that I already suffered through so you don't have to. You're welcome.

NO SHIT, THERE I WAS...
WELL, NOW YOU KNOW

Pain is a great teacher. Don't touch the stove because it's hot. Burn your hand once and you will not forget the lesson that this pain has caused you. Don't use 100 MPH Tape as a sock (see Shut Up and Ruck for this war story). The pain of that degloving injury will never be forgotten. I know exactly how fast I can run 1,000 meters with a 100 pound ruck. I know this, because the pain of that experience has so traumatized me that I will never forget.

During the Qualification Course, I had the great fortune of learning Small Unit Tactics under the tutelage of the great Paul Lefavor. He was a harsh critic of poor performance, but he was an incredible teacher and when he later found God (he went on to earn his Doctorate in Divinity) he proved to be an exceptional human being. But not the day he taught my team a valuable lesson. He was an absolute demon that day.

Paul had the habit of proving his manhood by making his teams carry the heaviest rucks. I recall that our rucks were so massive that we physically could not fit the requisite amount of MREs into the rucks themselves, and we were forced to string them up on the outside of the ruck with 550 cord. They looked like reactive armor plates and were an indication of how overburdened our loads were. But Paul was unfazed by our aching backs and shoulders.

We were out at Camp Mackall doing the Cadre-led terrain walk one day. Those that have endured this little exercise know that it is mostly used as a technique to smoke the shit out of students. You essentially run across the training area as fast as you can while the Cadre point out various terrain features as they whiz by your sprinting formation. At the very end of our terrain walk, Paul formed us up on a road. We were truly smoked. He asked us, way too nonchalantly, how fast we thought we could make a 1,000 meter movement. We were well drilled and responded that we thought we could do it in about 10 minutes.

That's a reasonably fast pace, but heavily burdened as we were, and as fatigued as we were, it would still prove a challenge. But it was a challenge that Paul felt we could overcome. He told us that we had 5 minutes to complete the movement. Make this 1,000 meter movement in 5 minutes or less, or we would be walking back to the barracks that night. No truck transport, an extra 10 mile ruck, and not an idle threat. Paul didn't make idle threats. Ever.

So he started his timer and took off like a bat out of hell. We chased him down in what must've looked like a scene from a movie. Paul, out front with a ruck full of pillows (probably) and us, rambling behind him in a full blown panic. Armor plated rucks and snot, spit, and sweat flinging everywhere. That may have been 5 of the hardest minutes that I've ever endured. I think that I may have blacked out at some point and pierced the temporal plane.

But I made it. We all made it. 1,000 meters in 5 minutes flat, with 100 pounds on my back. As we caught our breath, regained our composure, and gathered ourselves at the finish line Paul made the lesson clear. "Well, now you know that if you only have 5 minutes to exfil and you're 1,000 meters from the LZ, you won't miss your ride."

Fuck you, Paul. You're right, but fuck you.

10

THE END POINT

This brings us to the conclusion of our land navigation learning route. Let's review how we got here and talk about what's next. It has not been an insignificant process, so it deserves some reflection. And even if you retained every nugget of information that we presented, you still need more work to become a competent navigator. A couple of thematic trends that you should note is that everything that we covered is pretty basic. If you had zero land navigation exposure prior to picking up this book, then there was probably a good bit of confusion. That's totally normal and to be expected. If you had some exposure to land nav you probably already knew most of this stuff. This reinforces the declaration that there really isn't such a thing as advanced land navigation. Know where you are, know where you are going, and know how you are going to get there. Start point, route selection, end point. Distance and direction. Terrain association. Basics build champions.

We started to build our basics with a review of the map. We focused on military maps because that will constitute 99% of your land navigation experience. But you should familiarize yourself with other maps. USGS maps are fairly ubiquitous and follow the same design principles, particularly with regards to terrain, relief, and contour lines, so they are excellent training tools. Road maps are less helpful because they, for obvious reasons, focus on roads. But they still help the user build spatial awareness and orientation. Digital mediums, like satellite imagery and interactive GPS maps, can be helpful but if this is all you use to train with then you'll be disappointed, and likely confused, when you transition to a military map. Finally, you might consider a hand drawn map…hear me out.

A helpful tool in building map awareness is to do a detailed map study of a military map and then recreate that map from memory. For example, take a 10-kilometer by 10-kilometer area on a military map. Take 10 minutes and study it carefully. Then try to draw that area by hand from memory and see what you produce. After a few iterations you will start to see that you end up, by default, focusing on the key terrain to base your drawing on. This is helpful in that now you start to understand what key terrain really is. If you find yourself drawing roads and bowling alleys, then guess what the key navigable features are in your area? If you find yourself drawing rivers and mountain ranges, then guess what the key navigable features are in that area. I've actually done navigation practical exercises where we did just this...navigate from a hand drawn map drawn from memory.

After we got ourselves familiar with maps we dove right into grid systems. At first glance, the grid systems seem almost overwhelming, but if you start low and slow, like we did, then it's not too complicated. A simple quadrant, find the X, a bigger quadrant, find the X, and finally the kids carpet map and find the bank. We are just following this orderly, predictable, and repeatable method. A little context with latitude and longitude and a little detour for UTM. But we stay grounded in the MGRS. This leads us logically to plotting grid coordinates. You gotta do something with all of those grid lines! So we learned to plot grid coordinates and how to use a protractor. And we kept using that protractor for plotting azimuths. We started with maps and transitioned into how to use the map for finding direction and locations.

We took this to the next logical step...you have to do something now that you can determine locations with grid coordinates and direction with grid azimuths...and we learned to calculate distance. Straight line distance is pretty easy, especially with the bar graph. It gets a little trickier with curved measurements but as long as you take your time and make enough tick marks, you can get an accurate curved-line distance. And now we're really starting to make progress. In land navigation trifecta of know where you are, know where you're going, and know how you're going to get there we have all of the tools that we need. We haven't really talked about terrain and elevation and contours, but we have all of the baseline tools. Again, nothing advanced.

Having established the basics, we then took a little detour for intersection, resection, and modified resection. I want to stress again how even though there is so much of the standard land navigation foundations that you simply won't see during SFAS, they are still important. It's why they are foundational. Intersection, resection, and modified resection are obvious ones, but so are the five major, three minor, and two supplementary terrain features. You will only see a spur and draw at Camp Mackall. You will probably see way too much spur and draw so it will make up for the lack of cliff, valley, and saddle. But even though you won't use them, it's important that you have these tools in your toolbox. We aren't just preparing for SFAS, we are learning land navigation because knowing how to land navigate is a life skill.

And then we took a leap of faith and we started to plot routes. We started putting it all together with distance and direction, with terrain association and dead reckoning, and with checkpoints and attack points. We learned that time management must be deliberate and is absolutely critical during any assessment and it goes hand and hand with cognitive load management. If I had to prioritize the most critical components to successful land navigation, route planning would be the close number two to physical fitness as number one. If you are physically fit, then everything else becomes exponentially easier, including route planning. Along that route that we planned we learned about the importance of handrails, getting un-lost, and RV procedures. We are off to the races now.

We dabbled in night navigation with a foray into biology and science. The eye, the most critical sensor available to the navigator, is a complex structure and we learned how we can manipulate its usage with nutrition, hydration, and scanning techniques. There's not a lot that we can do, but we can do something and even a little something can go a long way. We learned little more about the compass and pace count and drift as well. You must learn to move well at night, and while the foundations and principles of night navigation are fundamentally identical to day navigation, it can be an entirely different beast. So we reiterated the tell, show, do methodology yet again. You have to get out and practice this stuff.

We wrapped up the "academic' portion, but we still had lots to learn with Always Look Cool. We talked again about the difference between terrain association and dead reckoning and learning when to use each method. We talked again about the compass and its nomenclature and how it works. We talked again about pace count and search procedures. We re-reviewed managing cognitive load. Do you see where I'm gong with this? We covered the same topics, several ways, with different contexts and applications. There is a reason for this, and its important.

There is no such thing as advanced land navigation. It doesn't exist. Land navigation is basic stuff. Know where you are (start point), know where you are going (end point), and know how you're going to get there (your route). Your route is determined by distance and direction. There are two ways to measure distance – pace count and time. There are two ways to measure direction – azimuth (dead reckoning) and terrain association. You take these basic principles and apply some discipline with good fitness, good cognitive load management, and practice and before too long you'll have achieved the pinnacle of human achievement. You'll be able to go anywhere in the world, whether that's behind enemy lines in a sweltering jungle and a massive ruck on your back, or to a local amusement park with a stroller and a diaper bag, and be confident that you will...Never Get Lost.

De Oppresso Liber

TELL, SHOW, DO

Step 1 is complete. You read the book. We **told** you. We dabbled a little in showing you, but it's tough to learn a tactile skill in a non-tactile modality. You should read this book again. Learn the terminology, become intimately familiar with the concepts, and commit to getting as fit as you can.

Step 2 is to **show** you. We have a comprehensive video library of all the skills that we told you about and we will endeavor to keep that library both updated and ever-growing. We'll cover the land navigation stuff, but we'll also cover gear and rucking and whatever other Green Beret and Green Beret adjacent stuff we think will serve aspiring Candidates and hobbyists alike.

Step 3 is to **do**. Find a skilled and reputable instructor and get out in the woods and practice. Find a local event or come to one of our Advanced Land Navigation Muster events if you can. You'll put all of this information into practice, you'll hear Green Beret war stories firsthand, and you'll get baptized at the infamous Scuba Road. All are welcome.

Visit TFVooDoo.com for all of our premium content, links to merchandise, books, and events, and for more world-class shit talking.

Ruck Up or Shut Up!

ALSO BY DAVID WALTON

Ruck Up or Shut Up is THE comprehensive guide to Special Forces Assessment and Selection, the gateway barrier to entry to the US Army Special Forces...the storied

Green Berets. This book is more than just a manual to get selected, this book is a prescriptive account of SFAS and a descriptive account of the culture, legend, and lore that surround SFAS.

Not everyone who reads this book will go to SFAS. There is still much to learn within these pages. There are great stories and lessons to be learned for just about anyone. For those who are already in the pipeline for Selection, these pages are a manual for what you can expect at Camp Mackall. For those on the path but not yet committed, you will likely find words of motivation and confirmation that will help you to commit. For those that are searching for something more, know that you have found it.

- Discover what makes Special Forces special
- Know what happens week by week at SFAS
- Identify how to build hard feet and ruck like a pro
- Learn why The Sandman never sleeps
- Recognize the best training methodologies
- Understand what gets you selected...and why

This book is for the hopeful few who will dare to try. You may not know what a Green Beret really does or precisely what it takes to earn the coveted title, but you feel the intense need to test your mettle. The good news is that if you are reading this book then you are headed in the right direction. Keep reading.

Then ruck up, we have work to do.

This is the ultimate program for Special Forces Assessment and Selection prep. Along with Ruck Up Or Shut Up, this guide provides everything needed to face the harshest assessment environment imaginable. SFAS is the ancestral proving grounds of the legendary Green Berets, and it requires aspiring operators to perform at the absolute edges of human performance. Preparation is the key to performance.

Shut Up and Ruck gives candidates inside access to top performance coaches in every assessed domain.

- 8 months of custom daily prep workouts
- See the science behind the process
- Tailored strength training for maximum gains
- Develop functional mobility to stay in the fight
- Build cardio for endless miles and the fastest times
- Ruck like a pro and avoid injury
- Fuel your prep with complete performance nutrition
- Learn the secrets to perfect sleep and total recovery
- Master your mental prep and resilience

Everything from the exercise science behind the big six lifts to the evidence based literature for the best way to build rucking performance. We cover the physiological realities of Zone 2 cardio and speed workouts. We discuss performance nutrition and supplementation, mobility and flexibility, injury prevention and skills building, cognition and resilience, and even the best methods to recover and sleep to support optimum performance. We even program complete daily workouts for a comprehensive 8-month SFAS prep with a custom performance journal to record and track your your data. We leave no stone unturned. No more excuses...

Everyone wants to be a Green Beret until it's time to do Green Beret sh*t.

Made in the USA
Coppell, TX
10 May 2025

49028860R00144